Satellite Meteorology

Third Edition

AF647415

Satellite Meteorology

Third Edition

R. R. Kelkar
M.Sc, Ph.D.

Former Director General of Meteorology,
India Meteorological Department.

Former ISRO Space Chair Professor,
Department of Atmospheric and Space Sciences,
Savitribai Phule Pune University.

BSP BS Publications™
An Imprint of **BSP Books Pvt.Ltd.**
4-4-309/316, Giriraj Lane,
Sultan Bazar, Hyderabad - 500 001
Telangana, India.

Satellite Meteorology, ***Third Edition***
by **R. R. Kelkar**

Copyright © 2026, *by Author*

Third Edition 2026
Second Edition 2016
First Edition 2006

No part of this book or parts thereof may be reproduced, stored in a retrieval system or transmitted in any language or by any means, electronic, mechanical, photocopying, recording or otherwise without the prior written permission of the authors.

Disclaimer: The authors and publishers have taken due care to provide authentic, reliable and up-to-date information related to the subject. However, neither the authors nor the publisher shall be responsible for any liability for any damage caused because of the use of this book. The respective users must check the accuracy from other sources too.

Published by:

BSP BS Publications™

An Imprint of **BSP Books Pvt.Ltd.**
4-4-309/316, Giriraj Lane,
Sultan Bazar, Hyderabad - 500 001, Telangana, India.
Phone: 040 - 23445688
e-mail: info@bspbooks.net
www.bspbooks.net

ISBN: 978-93-47311-18-5 (Paperback)

This book is dedicated to the memory of
my wife

MONICA

Preface to the Third Edition

Satellite meteorology made a humble beginning in April 1960 with the launch of the U.S. satellite TIROS-1. Today, there are as many as 18 geostationary meteorological satellites and 9 polar orbiting satellites of 7 countries and agencies, watching the earth's weather. Several other satellites in various types of orbits carrying specialized payloads are monitoring specific parameters of the earth's environment and climate. In July this year, India jointly with the U.S., has successfully launched the NISAR satellite which is said to be the most expensive environmental satellite built so far.

This rapid growth of satellite meteorology is the result of a complex interplay of scientific thinking, technological innovation and optimum use of resources. The third edition of *Satellite Meteorology* begins with the fundamentals as before, but its later chapters showcase the impressive achievements of missions that have been accomplished. Much of the increased accuracy of weather forecasts in recent years is attributable to the assimilation of satellite radiances and data into the global numerical models.

The last two chapters describe how satellite meteorology is now poised to spread across new domains. The trend now is towards integrated remote sensing that would provide accurate, long-term and consistent data about radiative forcing, greenhouse gas concentrations, aerosol characteristics, cloud properties, cyclone structure, and their interactions, all from a single satellite. In parallel, large constellations of small satellites are envisaged to be launched for achieving repetivity on a scale which was so far considered impossible. It is also notable that several private enterprises are coming forward to support such ventures.

The age of AI has already dawned and future satellites are expected to be capable of processing data on board without losing precious time in transmitting it to a ground station. Satellites which serve as observing platforms will, with AI, soon begin functioning instead as intelligent systems. This would be critical for real time forecasting of severe storms, lightning strikes, wildfires and flash floods. Satellite meteorology will no longer remain confined to measurement of cloud height, temperature, humidity or wind. It will offer deeper insights into climate change, enable faster decision making in disaster situations, and help to achieve sustainable development.

While foreseeing all these happenings in the not too distant future, I present the fully updated third edition of *Satellite Meteorology* to my readers, old and new. I sincerely hope that it will be as well received as the earlier editions by meteorologists, climate scientists, teachers, students, and all those who are working towards making our planet a better place to live.

R. R. Kelkar

Preface to the Second Edition

Our world has changed in the last ten years. In satellite meteorology, the change has been phenomenal. Many satellites which were in orbit when I wrote the first edition of this book in 2006, are no longer there. Some instruments which were then on the design board are now in space. Concepts which then seemed like a flight of imagination are now close to realization.

India has sent space missions to the moon and Mars, and it has made a quantum jump in satellite meteorology with the launch of INSAT-3D in 2013. This is a dedicated meteorological satellite with a 6-channel imager and a 19-channel sounder in geostationary orbit. India is now the only country in the world besides the U.S. to have a geostationary satellite with a sounding capability.

The last decade has witnessed a significant improvement in the accuracy of short and medium range weather forecasting the world over, particularly in the tropics. Numerical weather prediction models seem to be taking over from synoptic meteorologists and may even be doing better. With the support of high power computers, numerical models have indeed become sophisticated and highly capable. However, it is undeniable that their recent success has been largely due to the real time assimilation of satellite data and products.

Against the backdrop of these developments, I was feeling that a revision of *Satellite Meteorology* had become overdue. A few months back, I received a reminder to that effect from my publishers and I started working on the revision. The second edition retains the basic structure and style of my earlier book, but the updated content reflects more realistically the state of art in this ever-evolving field of science and technology. It incorporates the most recent factual and technical information, research results and references to the latest publications.

In 2003 I retired as the Director General of India Meteorological Department, New Delhi. In 2008 my term as ISRO Chair Professor, University of Pune, also came to an end. Although I no longer hold any official or academic position, my interest in satellite meteorology remains unabated. I am glad to present the second edition of *Satellite Meteorology* to my readers and I hope that it will prove to be as useful and popular as the first.

Pune, 23 March 2016 R. R. Kelkar

Preface to the First Edition

When I was a young recruit of the India Meteorological Department, and working at the Institute of Tropical Meteorology in Pune, I was once sent to Mumbai for getting the customs clearance of a shipment that had arrived from the U. S. It was a heavy wooden crate that contained data from a single pass over India by the U. S. satellite, TIROS-4. The data was on a large spool of magnetic tape, accompanied by several volumes of instruction manuals that explained how the tape was to be read. To my disappointment, our attempts to retrieve the data were unsuccessful, but my fascination for satellites and satellite meteorology did not end there.

In 1980, I was sent on a 10-week training programme in satellite image processing at the NOAA facilities in Washington DC. On my return, I joined the IMD team that was getting the Meteorological Data Utilization Centre into readiness for the launch of the first Indian National Satellite, INSAT-1A. From then onwards, until 1995, I had the opportunity to work in almost every branch of satellite meteorology: software development, image processing, product derivation, algorithm development, hardware design, ground system integration, earth station technology, radiometer design, satellite orbit design, and finally, overall project implementation and management.

After my retirement in 2003 as Director General of IMD, I joined the University of Pune as ISRO Chair Professor and during 2005-06, I taught a 2-semester course in satellite meteorology for students of M. Sc. (Atmospheric Sciences). I found that there was no single book available that I could recommend to the students, and some topics that I was teaching were not found in books at all. This is what motivated me to write my own book on satellite meteorology.

India is one of the few countries of the world that makes and launches its own satellites. India is also the land of the monsoons. It is just natural that my book has an emphasis on Indian satellites and Indian weather and climate. I am sure that this book on satellite meteorology will not only be of help to university students, but also be a source of information to operational meteorologists, research workers and all others who have more than a casual interest in the subject.

Pune, November 2006 R. R. Kelkar

Acknowledgements

When I wrote the first edition of my book, the India Meteorological Department (IMD) had very kindly provided me a large number of INSAT and Kalpana-1 satellite images and given me permission to reproduce them in the book. I am using many of them here for which I am expressing once again my sincere thanks.

Several recent images have been added in the third edition. These have been taken from the internet web sites in the public domain, and the source URL is mentioned.

Wherever significant scientific and technical information has been quoted from published work, the authors have been duly acknowledged in the text and publication details given in the list of references.

Finally, I thank my publishers, Mr. Nikhil Shah, Mr. Anil Shah and Mr. Naresh Davergave of BSP Books Pvt.Ltd., for their continuing interest in my books and for motivating me to keep writing.

R. R. Kelkar

List of Acronyms and Abbreviations

AATSR	Advanced Along Track Scanning Radiometer
ABI	Advanced Baseline Imager
ACRIM	Active Cavity Radiometer Irradiance Monitor
ACRIMSAT	ACRIM Satellite
ADCS	Advanced Data Collection System
ADEOS	Advanced Earth Observation Satellite
AISMR	All-India Summer Monsoon Rainfall
ALtiKa	ALtImeter Ka band
AMR	Advanced Microwave Radiometer
AMV	Atmospheric Motion Vectors
AMSU	Advanced Microwave Sounding Unit
AOD	Aerosol Optical Depth
AOS	Atmospheric Ocean System
APT	Automatic Picture Transmission
ASCAT	Advanced Scatterometer
ATSR	Along Track Scanning Radiometer
AVHRR	Advanced Very High Resolution Radiometer
CALIPSO	Cloud-Aerosol Lidar and Infrared Pathfinder Satellite Observation
CALIOP	Cloud-Aerosol Lidar with Orthogonal Polarization
CCD	Charge Coupled Devices
CDO	Central Dense Overcast
CHAMP	CHAllenging Minisatellite Payload
CIMSS	Cooperative Institute for Meteorological Satellite Studies
CLIPER	Climatology and Persistence
CMW	Cloud Motion Wind
COSMIC	Constellation Observing System for Meteorology, Ionosphere and Climate
CPR	CloudSat Precipitation Radar
CSC	Cloud System Centre
CI	Current Intensity
DIAL	Differential Absorption Lidar

DMSP	Defense Meteorological Satellite Programme
DNB	Day/Night Band
DORIS	Doppler Orbitography and Radio-positioning Integrated by Satellite
DPR	Dual-frequency Precipitation Radar
DWL	Doppler Wind Lidar
DWSS	Defense Weather Satellite System
ECMWF	European Centre for Medium-range Weather Forecasts
ENSO	El Nino Southern Oscillation
ENVISAT	Environmental Satellite
ERBE	Earth Radiation Budget Experiment
ERS	European Remote Sensing Satellite
ESA	European Space Agency
ESMR	Electrically Scanning Microwave Radiometer
ESSA	Environmental Science Services Administration
EUMETSAT	European Organization for the Exploitation of Meteorological Satellites
FCI	Flexible Combined Imager
FY	Feng-Yun
GCOM-W1	Global Change Observation Mission - Water 1
GERB	Geostationary Earth Radiation Budget
GHGSat	GreenHouse Gas Satellite
GIFTS	Geostationary Imaging Fourier Transform Spectrometer
GISAT	Geostationary High Resolution Imager Satellite
GLM	GEO Lightning Mapper
GMI	GPM Microwave Imager
GMS	Japanese Geostationary Meteorological Satellite (Himawari)
GODAE	Global Ocean Data Assimilation Experiment
GOES	Geostationary Operational Environmental Satellite
GOME	Global Ozone Monitoring Experiment
GOMS	Geosynchronous Operational Meteorological Satellite
GPCP	Global Precipitation Climatology Project
GPI	GOES Precipitation Index
GPM	Global Precipitation Measurement mission

GPROF	Goddard Profiling Algorithm
GPS	Global Positioning System
GRAS	Global Navigation Satellite System Receiver for Atmospheric Sounding
GSPDC	Geostationary Satellite Data Processing Centre
HES	Hyperspectral Environmental Suite
HIRS	High Resolution Infra-red Sounder
HRC	Highly Reflective Clouds
HSRL	High Spectral Resolution Lidar
IASI	Infrared Atmospheric Sounding Interferometer
IFOV	Instantaneous Field Of View
IGeo Lab	International Geostationary Laboratory
IMD	India Meteorological Department
INCOIS	Indian National Centre for Ocean Information Services
INSAT	Indian National Satellite
IR	Infra-red
ISCCP	International Satellite Cloud Climatology Project
ISRO	Indian Space Research Organisation
ITCZ	Intra-Tropical Convergence Zone
ITML	Interaction between Tropical and Mid-Latitude Systems
JAXA	Japan Aerospace Exploration Agency
JPSS	Joint Polar Satellite System
LEO	Low Earth Orbit
LFPA	Large Area Focal Plane detector Array
LIS	Lightning Imaging Sensor
LLJ	Low Level Jet
LRA	Laser Retroreflector Array
LST	Land Surface Temperature
LSTD	Diurnally averaged Land Surface Temperature
MADRAS	Microwave Analysis and Detection of Rain and Atmospheric Structures
MetOp	Meteorological Operational Satellite
METSAT	Meteorological Satellite
MHS	Microwave Humidity Sounder

MIRAS	Microwave Imaging Radiometer with Aperture Synthesis
MODIS	Moderate Resolution Imaging Spectroradiometer
MOS	Modular Optoelectronic Scanner
MSG	Meteosat Second Generation
MSMR	Multi-Spectral Microwave Radiometer
MST	Mesosphere Stratosphere Troposphere Radar
MSU	Microwave Sounding Unit
MTC	Mid-Tropospheric Cyclone
MTG	Meteosat Third Generation
MTSAT	Multi-purpose Transportation Satellite
MWIR	Mid-wave infra-red
NASA	National Aeronautics and Space Administration
NCMRWF	National Centre for Medium Range Weather Forecasting
NDVI	Normalized Difference Vegetation Index
NIR	Near infra-red
NISAR	Nasa Isro Synthetic Aperture Radar satellite
NOAA	National Oceanic and Atmospheric Administration
NPOESS	National Polar-orbiting Operational Environmental Satellite System
NPP	NPOESS Preparatory Project
NSCAT	NASA Scatterometer
OCM	Ocean Colour Monitor
OCO	Orbiting Carbon Observatory
ODT	Objective Dvorak's Technique
OLR	Outgoing Longwave Radiation
OLS	Operational Linescan System
OSTM	Ocean Surface Topography Mission
OTD	Optical Transient Detector
PACE	Plankton, Aerosol, Cloud, and ocean Ecosystem mission
PARASOL	Polarization and Anisotropy of Reflectances for Atmospheric Sciences coupled with Observations from a Lidar
PFZ	Potential Fishing Zone
POLDER	POLarization and Directionality of the Earth's Reflectances

POES	Polar Orbiting Operational Environmental Satellite
PROBA	Project for On-Board Autonomy
PSLV	Polar Satellite Launch Vehicle
QBO	Quasi-Biennial Oscillation
QPE	Quantitative Precipitation Estimate
QuikSCAT	Quick Scatterometer
RTE	Radiative Transfer Equation
SAGE	Stratospheric Aerosol and Gas Experiment
SAMIR	Satellite Microwave Radiometer
SAPHIR	Sounder for Atmospheric Profiling of Humidity in the Inter-tropics by Radiometry
SAR	Search And Rescue
SAR	Synthetic Aperture Radar
SARAL	Satellite with Argos and ALtiKa
SBDART	Santa Barbara DISORT Atmospheric Radiative Transfer
ScaRaB	Scanner for Radiation Budget Measurement
SCIAMACHY	Scanning Imaging Absorption Spectrometer for Atmospheric CHartographY
SeaWiFS	Seastar Wide Field of View Sensor
SEM	Space Environment Monitor
SHET	Southern Hemispheric Equatorial Trough
SIM	Spectral Irradiance Monitor
SMAP	Soil Moisture Active Passive Satellite
SMMR	Scanning Multichannel Microwave Radiometer
SMOS	Soil Moisture and Ocean Salinity Satellite
SMS	Synchronous Meteorological Satellite
SOLSTICE	SOLar STellar Irradiance Comparison Experiment
SORCE	Solar Radiation and Climate Experiment
SSM/I	Special Sensor Microwave Imager
SSM/T	Special Sensor Microwave Temperature Profiler
SSM/T2	Special Sensor Microwave Humidity Profiler
SST	Sea Surface Temperature
SSU	Stratospheric Sounding Unit
STJ	Sub-Tropical Jet Stream

SWIR	Short Wave infra-red
TBB	TriBand Beacon Transmitter
TEJ	Tropical Easterly Jet
TIM	Total Irradiance Monitor
TIP	Tiny Ionospheric Photometer
TIR	Thermal infra-red
TIROS	Television and Infra-Red Observation Satellite
TIROS-N	TIROS Next Generation satellite
TMI	TRMM Microwave Imager
TMPA	TRMM-based Multi-satellite Precipitation Analysis
TOMS	Total Ozone Mapping Spectrometer
TOPEX	Topography Experiment
TOVS	TIROS Operational Vertical Sounder
TRMM	Tropical Rainfall Measuring Mission
TROPICS	Time-Resolved Observations of Precipitation structure and storm Intensity with a Constellation of Smallsats
UARS	Upper Atmosphere Research Satellite
UTWV	Upper Tropospheric Water Vapour
UV	Ultra-violet
VAS	VISSR Atmospheric Sounder
VHRR	Very High Resolution Radiometer
VIIRS	Visible-Infrared Imager-Radiometer Suite
VIRS	Visible and InfraRed Scanning Radiometer
VIS	Visible
VISSR	Visible Infrared Spin-Scan Radiometer
WALES	Water Vapour Lidar Experiment in Space
WCRP	World Climate Research Programme
WD	Western Disturbance
WOCE	World Ocean Circulation Experiment
WRF	Weather Research and Forecasting Model
WV	Water Vapour
XPS	X-Ray and U-V Photometer System

Contents

2. Weather Systems Observed in Satellite Imagery

Chapter 1

Fundamentals

Seventy years ago, satellite meteorology did not exist except perhaps in science fiction. Today, the latest satellite images of the world's weather and animated sequences of tropical storms are being shown on our television sets and available on our mobiles. The fascinating origin of this new branch of meteorology, and its phenomenal growth, have indeed had a touch of fantasy. What satellite meteorology happens to be today, is the result of an interplay of science on one hand, and the technology of satellites, computers and communications on the other. Limitations of technology have been overcome by scientific ingenuity, and the requirements of science have driven technology to the cutting edge.

After the successful launch of the first weather satellite TIROS-1 in 1960 and the growth of satellite coverage of the earth's atmosphere and oceans within just a decade afterwards, meteorologists were in fact overwhelmed by the new satellite data that became available to them. Prior to the satellite era, meteorologists had laid the greatest emphasis on atmospheric pressure. Lows and highs showing up in synoptic isobaric analysis were of their main interest and many other weather elements including cloud cover, although observed, were not analyzed on a synoptic scale in a similar manner. With the availability of satellite images, however, the accent shifted to observing and examining clouds and cloud patterns in the imagery, from which the state of the atmosphere could be directly observed or inferred. Images received from geostationary and orbiting satellites together revealed the presence of a wide spectrum of atmospheric phenomena across individual cumulus cells, thunderstorms, tropical cyclones and jet streams, just at a glance. So much so, that ground weather observation stations began to face the risk of redundancy and closure. Competition from satellite imagery forced many national meteorological services to take a re-look at their network configurations and trim them, particularly the upper air stations which are expensive to operate and maintain.

Whether satellites can give us all that we need for weather analysis and forecasting has been a subject of debate, what is unquestionable is the fact that satellites have been giving us more and more than we ever anticipated. Satellites are no longer playing just a supportive role in meteorology but they are becoming independent and intelligent systems that would manage climate change, disaster response and sustainable development.

Whatever be the scenario, present or future, the fundamental principles of satellite meteorology do not change with time and they are discussed in the following sections of this chapter.

1.1 Principles of Meteorological Remote Sensing

Remote sensing has been defined in various ways, but it is basically the process of observing an object without going near it or touching it, and that too in wavelengths that the human eye cannot perceive. Remote sensing of the earth and its atmosphere was a novel idea in the mid-20th century and was done primarily by taking photographs from aircrafts. Soon thereafter it developed into a satellite-based technique for measurement of radiation from the sun and the earth. Remote sensing satellites were launched for monitoring the earth's natural resources and meteorological satellites for observing the atmosphere.

1.1.1 Electromagnetic Spectrum

By the term spectrum, we traditionally mean the seven colours of visible light, such as those seen in a rainbow. Nowadays, the term has come to be associated more with mobile communications along with terms like 4G or 5G. In scientific parlance, however, it refers to the entire range of wavelength or frequency of electromagnetic radiation, visible light or the microwave region being just small parts of it (Table 1.1.1.1). The characteristic spectrum of a given object is the pattern of electromagnetic radiation that it absorbs, transmits and emits.

The product of the wavelength λ and frequency ν of electromagnetic waves is equal to c, the velocity of light. The associated energy $E = h.\nu = h.c / \lambda$ where h is Planck's constant. This means that as the wavelength of electromagnetic radiation increases, its frequency decreases and the associated energy gets reduced.

The region of the electromagnetic spectrum with which we are most concerned in real life is the region of visible light, to which the human eye is very sensitive and in which the sun and stars emit the strongest radiation. In recent times, we are getting familiar with other wavelength regions as FM radio stations, mobile phones, satellite television or microwave ovens become more and more a part of our daily life.

Table 1.1.1.1 Electromagnetic Spectrum.

<table>
<tr><th>Wavelength</th><th></th><th>Wavelength</th><th></th></tr>
<tr><td>10^{-6} nm</td><td rowspan="5">Gamma Rays
(MeV)</td><td>1 mm</td><td>Millimetre Waves (mm)</td></tr>
<tr><td>10^{-5} nm</td><td>1 cm</td><td rowspan="3">Microwaves
(cm, GHz)</td></tr>
<tr><td>10^{-4} nm</td><td>10 cm</td></tr>
<tr><td>10^{-3} nm</td><td>1 m</td></tr>
<tr><td>10^{-2} nm</td><td>10 m</td><td rowspan="8">Radio Waves
(MHz, kHz)</td></tr>
<tr><td>10^{-1} nm</td><td rowspan="3">X-Rays (Å)</td><td>100 m</td></tr>
<tr><td>1 nm</td><td>1 km</td></tr>
<tr><td>10 nm</td><td>10 km</td></tr>
<tr><td>100 nm</td><td rowspan="2">Ultra-Violet (nm),
Visible, Near Infra-Red(μ)</td><td>100 km</td></tr>
<tr><td>1 μ</td><td>10^3 km</td></tr>
<tr><td>10 μ</td><td>Thermal Infra-Red (μ)</td><td>10^4 km</td></tr>
<tr><td>100 μ</td><td>Far Infra-Red (μ)</td><td>10^5 km</td></tr>
</table>

Note: 1 nm (nanometre) = 10^{-9} m and 1 μ (micrometre or micron) = 10^{-6} m

Table 1.1.1.2 Wavelength Range of Visible Colours.

Colour	**Wavelength**	
	(nm)	**(μ)**
Violet	380-430	0.38-0.43
Indigo	430-500	0.43-0.50
Blue	500-520	0.50-0.52
Green	520-565	0.52-0.565
Yellow	565-590	0.565-0.59
Orange	590-625	0.59-0.625
Red	625-740	0.625-0.740

The seven colours of the visible spectrum are identified by their wavelengths (Table 1.1.1.2). Radiation of wavelengths shorter than violet is called ultra-violet (UV) radiation. This has very high energy that can break chemical bonds, ionize molecules, damage skin cells or cause cancer. However, most of the UV radiation coming from the sun is absorbed by the layer of atmospheric ozone which resides in the stratosphere, and shields life on earth from its harmful effects.

X-rays have wavelengths that are even shorter than UV, which are expressed in Å (Angstrom Units or 10^{-10} m). Gamma rays have wavelengths that could be as short as 10^{-15} m and it is more convenient to express their magnitude in terms of their energy levels which are of the order of KeV (Kilo electron volts) or MeV (Million electron Volts). X-rays and gamma rays have great penetration power and have applications in astronomy, radioactivity, medicine, and other fields.

Towards the other end of the visible spectrum, radiation which has wavelength higher than red is called infra-red (IR). The IR region of the spectrum can be further sub-divided into near (NIR), short-wave (SWIR), middle (MIR), and thermal (TIR) with increasing wavelength.

Radiation with still longer wavelengths is known by different names such as millimetre waves, microwaves and radio waves. These again are further classified with respect to their frequency as given in Tables 1.1.1.3 and 1.1.1.4.

Table 1.1.1.3 Nomenclature of Microwave and Radio Wave Frequencies.

Abbreviation	Full Form	Frequency	Wavelength
EHF	Extremely high frequency (Microwaves)	30-300 GHz	1 mm-1 cm
SHF	Super high frequency (Microwaves)	30-3 GHz	1 cm-10 cm
UHF	Ultra-high frequency	3 GHz-300 MHz	10 cm-1 m
VHF	Very high frequency	300-30 MHz	1 m-10 m
HF	High frequency	30-3 MHz	10 m-100 m
MF	Medium frequency	3 MHz-300 kHz	100 m-1 km
LF	Low frequency	300-30 kHz	1-10 km
VLF	Very low frequency	30-3 kHz	10-100 km
VF	Voice frequency	3 kHz-300 Hz	100-10^3 km
ELF	Extremely low frequency	300-30 Hz	10^3-10^4 km

Table 1.1.1.4 Microwave Bands.

Band	Wavelength	Frequency
mm-Band	1-7.5 mm	40-300 GHz
Ku-K-Ka- Band	0.75-2.5 cm	12-40 GHz
X-Band	2.5-4 cm	8-12 GHz
C-Band	4-8 cm	4-8 GHz
S-Band	8-15 cm	2-4 GHz
L-Band	15-30 cm	1-2 GHz

1.1.2 Absorption, Emission, Reflection and Scattering

It is quite a paradox that the radiative and thermal equilibrium of the earth-atmosphere system is not controlled by the two major constituents of the atmosphere which are nitrogen and oxygen, but by a few of the numerous miscellaneous gases that all put together make up for just 1% of its volume. The behaviour of gases like water vapour, carbon dioxide, ozone, methane and other trace gases, and also particulate matter floating in the atmosphere, is what alters the radiation from the sun traversing the atmosphere and the radiation returned to space by the earth. Since satellite remote sensing is essentially the measurement of the returned radiation, it is important to know how the gases in the atmosphere influence the radiative processes in the atmosphere.

Atoms and molecules in a gas have electronic, rotational or vibrational energy, and absorption or emission of radiation takes place when there is a transition from one energy state to another. Absorption spectra of atoms, such as atomic oxygen and nitrogen, are associated with electronic transition and occur in the ultra-violet (UV) region of the electromagnetic spectrum. Tri-atomic molecules like those of water vapour, carbon dioxide and ozone, have additional rotational and vibrational transitions, which occur mainly in the infra-red (IR) region. In the visible (VIS) region, gases in the atmosphere account for very little absorption. The main absorption bands are those of three atmospheric gases: water vapour at 6.7 μ, carbon dioxide 15 μ and ozone 9.6 μ (1 μ or micron = 10^{-6} m). There are other minor absorption bands attributable to methane, nitrous oxide, and other gases.

Radiation from the sun gets reflected when it strikes a plane surface such as the ground or cloud tops and its direction gets altered. Depending upon the albedo or reflectivity of the surface, some part of the radiation will be reflected and the remaining amount will get absorbed by the medium or be transmitted through it. Snow and cumulonimbus cloud tops have high albedo values while the ocean surface reflects very little of the radiation falling on it.

Air molecules and suspended particles or aerosols scatter radiation in the VIS wavelengths. When the size of the scattering particles is small compared to the wavelength of the incident radiation, the scattering is said to be of the Rayleigh type. In Rayleigh scattering, the intensity is inversely proportional to the fourth power of the wavelength, and the distribution of scattered radiation intensity is symmetric in both the forward and backward directions. When the sizes of the scattering particles become comparable to the wavelength of the incident radiation, the scattering processes is said to be of the Mie type, in which the angular distribution of the scattered radiation intensity is complex and asymmetric. Rayleigh scattering by air molecules is what gives the blue colour to the sky, while Mie scattering by the larger sized particles and aerosols gives it a grayish appearance.

1.1.3 Black Body and Radiation Laws

Many fundamental laws governing the absorption and emission of electromagnetic radiation are commonly based on the concept of what is termed as a black body. This is largely a theoretical concept as an ideal or perfect black body can be said to be almost non-existent in reality. A black body is defined as an object that absorbs all radiation incident upon it, does not reflect any of it, and emits all energy at full efficiency for all wavelengths as per the following equation

$$B(\lambda,T) = 2hc^2\lambda^{-5} / (e^{hc/\lambda kT} - 1)$$

where B is the energy in w m^{-2} μ^{-1},
T is the temperature of the black body in °K,
λ is the wavelength in μ,
h is Planck's constant 6.625×10^{-27} erg sec,
k is Boltzmann constant 1.38×10^{-16} erg °K^{-1},
c is the velocity of light 3×10^{10} cm sec^{-1}

This relationship of the black body emission to its temperature is known as Planck's Law. For any given temperature, Planck's Law gives a distribution of emitted energy or a characteristic spectrum of electromagnetic radiation that peaks at a certain wavelength. This peak shifts to shorter wavelengths for higher temperatures and the area under the curve grows rapidly with increasing temperature.

Wien's Displacement Law describes the temperature dependence of the black body radiation curves derived from Planck's Law and it states that the wavelength λ_{max} at which the black body radiation is the maximum is inversely proportional to the temperature T of the black body (in °K), or

$$\lambda_{max} = c / T$$

where c is a constant whose value is 2898 μ °K.

Stefan-Boltzmann's Law gives the total energy E emitted at all wavelengths by the black body at temperature T,

$$E = \sigma T^4$$

where σ is the Stefan-Boltzmann constant which has a value of 5.6705×10^{-5} erg cm^{-2} sec^{-1} $°K^{-4}$.

E is in fact the area under the Planck's Law curve. Thus, as the temperature of a black body increases, the shift of the peak emission to shorter wavelengths is governed by Wien's Law while the increase in the height of the curve is explained by Stefan-Boltzmann's Law. This increase in E with temperature is not linear, since it varies with the fourth power of the temperature.

There is another important law called Kirchhoff's Law which states that the ratio ε of emitted radiation to absorbed radiation is the same for all black bodies at the same temperature. This law forms the basis for the definition of emissivity. The emissivity of a perfect black body is 1 and that of a perfect reflector is 0. Strictly speaking, the laws mentioned above are applicable only to a black body, but they are important in the real world as they can be applied as a close approximation to other bodies which have a very weak interaction with the surrounding environment and can be considered to be in a state of equilibrium. In the earth-atmosphere system, the oceanic surface and tall dense clouds can be regarded as acting very similar to a black body.

1.1.4 Solar and Terrestrial Radiation

The radiation laws described in the previous section help us to understand the nature of an object and identify some of its thermal properties by interpreting the pattern of radiation emitted by it in different wavelengths. The spectrum of the solar radiation received at the top of the earth's atmosphere matches very well the spectrum of a black body having a surface temperature of about 5700 °K. This is called solar radiation and it has a peak at about 0.5 μ (Figure 1.1.4.1). The radiation emitted by a black body at the same temperature as the average temperature of the earth's surface which is 283 °K, peaks at about 10 μ (Figure 1.1.4.2). We therefore have to deal with two different radiation regimes, the radiation received by the earth from the sun, called solar or shortwave radiation and the radiation returned to space by the earth-atmosphere system, called terrestrial or infra-red or longwave radiation. The intensity of terrestrial radiation is far less than that of solar radiation.

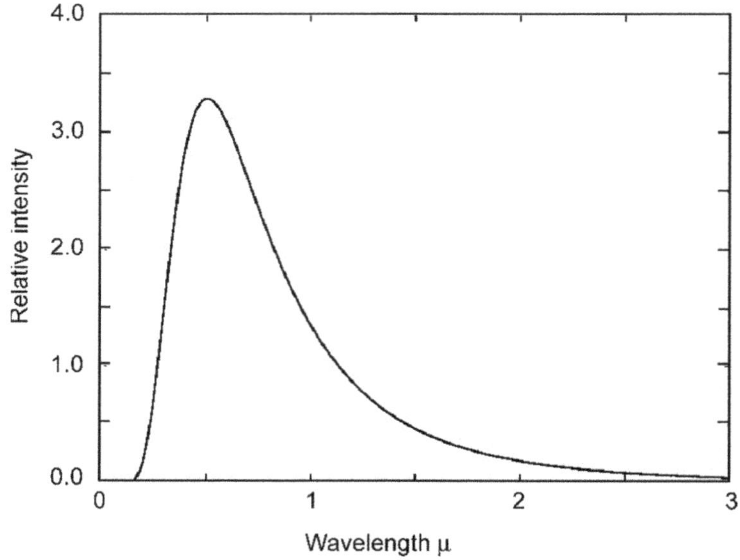

Figure 1.1.4.1 Variation of the relative intensity of black body radiation with wavelength (μ) at temperature 5700 °K, peaking at a wavelength of 0.5 μ in the visible region (Solar radiation).

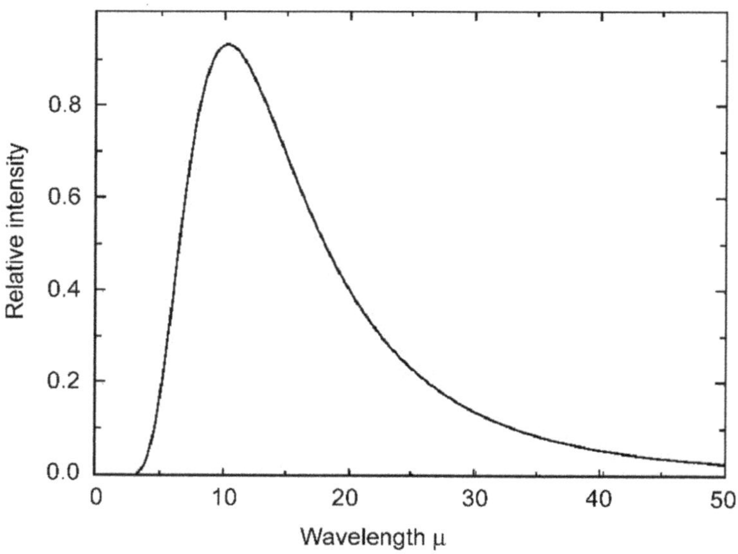

Figure 1.1.4.2 Variation of the relative intensity of black body radiation with wavelength (μ) at temperature 283 °K, peaking at a wavelength of 10 μ in the infrared region (Terrestrial radiation).

1.1.5 Sun-Earth-Atmosphere Radiation Budget

The solar constant is defined as the annual average solar radiation received outside the earth's atmosphere on a plane normal to the incident radiation at the mean sun-earth distance and has a value close to 2 cal cm^{-2} min^{-1} or 1370 W m^{-2}. The actual solar irradiance varies by 3-4% of this value during the year due to the eccentricity of earth's orbit about the sun.

If we consider the incoming solar radiation at the top of the atmosphere as made of 100 units, 30 units are reflected back to space (6 by the atmosphere, 20 by clouds and 4 by the earth's surface). 19 units are absorbed by the atmosphere (16 by gases and 3 by clouds). The remaining 51 units are absorbed by the earth's surface.

Out of these 51 units, 6 are lost to space directly and 45 are returned upwards and absorbed by the atmosphere and clouds (7 by convection and conduction, 23 by evaporation as latent heat and 15 by longwave radiation). The atmosphere and clouds have already absorbed 19 units from the solar radiation, making a total of 64 units which are returned to space as longwave radiation. The budget is thus balanced at the top of the atmosphere.

At the earth's surface and at any level in the atmosphere, the net radiation is the balance of four radiative fluxes, downward solar radiation, downward longwave radiation, upward solar radiation and upward longwave radiation. These can be measured with special instruments installed on the ground or flown on balloons as radiometersondes. The prime factors involved in the radiation budget of the earth-atmosphere system are the albedo or reflectance properties of land, ocean and cloud tops, scattering properties of aerosols, dust and particulate matter in the atmosphere, and vertical profiles of temperature and concentration of gases which absorb longwave radiation (water vapour, CO_2, ozone). Data on these variables if available can be used to compute the radiation budget components indirectly.

1.2 Satellite Orbits

The design of an optimum orbit around the earth for a meteorological satellite is a complex process. There are two main classical types of orbits, polar and geostationary, which are complementary to each other. However, over time, many other types of orbits have come into use or have been tried. The following sections describe the fundamental principles behind orbit design, technical considerations and the operational implications.

1.2.1 Gravitational and Astronomical Laws

There are certain classical laws that were originally formulated to explain the motion of planets in the solar system and their orbits around the sun. They are, however, very fundamental and general in nature and we now know that they are equally applicable to moons in the solar system or the orbits of artificial satellites placed around the earth and other planets.

Kepler's laws of motion state that: (i) A planet moves around the sun in an elliptical orbit, with the sun at one focus, (ii) The vector joining the sun's centre to the planet sweeps out equal areas in equal time, and (iii) The square of the period of revolution of the planet is proportional to the cube of its semi-major axis. As mentioned above, these laws also hold good for artificial satellites orbiting the earth.

Kepler's laws have to be considered in conjunction with another fundamental physical law. As per Newton's law of gravitation, the gravitational force between two bodies of mass m_1 and m_2 is proportional to the product of m_1 and m_2 and inversely proportional to the square of the distance r between them, or

$$F = G\, m_1\, m_2 / r^2$$

where G is the universal gravitational constant having a value of 6.67×10^{-8} $cm^3\, g^{-1}\, sec^{-2}$.

The forces acting on a satellite around the earth are the gravitational force and the centripetal force, which should balance for the satellite to attain a stable orbit. So we must have

$$m_s\, v^2 / R = G M_e m_s / R^2$$

where M_e and m_s are the masses of the earth and satellite respectively, R is the mean distance between them and v is the velocity of the satellite. The term m_s appears on both sides of the equation and can be cancelled out. We then have

$$v^2 = G M_e / R$$

For a circular orbit, the velocity v can be expressed as $2\, \pi\, R / T$ where T is the time period of revolution of the satellite, which is inversely related to its mean distance from the earth and is given by the equation

$$T^2 = 4\, \pi^2\, R^3 / G M_e$$

This means that the time period, speed and acceleration of an artificial satellite orbiting the earth are not dependent upon its mass. So theoretically speaking we can put into orbit as big a satellite as we wish, the only practical constraint being that of lifting it into space with the rockets that we have.

It needs to be clarified here that the centripetal force is not independent but derived from the continuous gravitation-related fall of the satellite. However, that does not mean that the gravitational force can on its own sustain the satellite stability. Orbits begin to decay slowly with time and the imbalance of forces may need to be occasionally corrected by firing on-board thrusters.

We can evaluate T for a given R as

$$T = (4\,\pi^2\,R^3\,/GM_e)^{1/2}$$

R can be expressed as $R_e + h$, where R_e is the radius of the earth and h is the height of the satellite above the earth. For the limiting case of h = 0, and using $M_e = 5.98 \times 10^{24}$ kg, and $R_e = 6370$ km, T works out to be 84 min. In reality of course it is not possible to have a satellite grazing the earth. For h = 200 km, T will work out to be 88 min. For h = 1000 km, T will be 105 min, and this is a popular choice for meteorological satellites.

If T is set at 24 hr, the height of the satellite will work out to be about 35,840 km above the earth's surface. Such an orbit is called geosynchronous as the satellite matches the angular velocity of the earth at this height.

1.2.2 Orbital Elements

As depicted in Figure 1.2.2.1, the satellite's orbit is specified by the following parameters (Kelkar et al 1980):

(a) a - the semi-major axis of the orbit (defines the size of the orbit)

(b) e - the eccentricity of the orbit (defines the shape of the orbit)

(c) I - the inclination of the orbit (with respect to the earth's equatorial plane)

(d) Ω - the right ascension of the ascending node (longitude of the north-bound equatorial crossing)

(e) ω - the argument of the perigee

(f) f - true anomaly

(g) T_e - epoch time

For a given orbit, all the above parameters are fixed, except f which specifies the position of a satellite at a given time.

The semi-major axis a of the ellipse is the maximum height of the satellite above the earth's surface. As per Kepler's third law, satellites move faster when they are close to the earth and slower when they are further away. The semi-major axis is thus an indirect measure of the average speed or mean motion of the satellite.

The satellite orbit is generally an ellipse. The eccentricity e is an indicator of the shape of the ellipse. A circular orbit is a special case of an ellipse when e is 0, and as e increases towards the limiting value of 1, the ellipse becomes more and more eccentric or elongated.

The point where the satellite is closest to the earth is called the perigee, and the point where it is farthest from the earth is called the apogee. The major axis of the ellipse is thus the line joining the perigee and the apogee. The argument of perigee ω helps to define the orientation of the orbital ellipse within the orbital plane. It is the angle measured at the centre of the earth from the ascending node to the perigee.

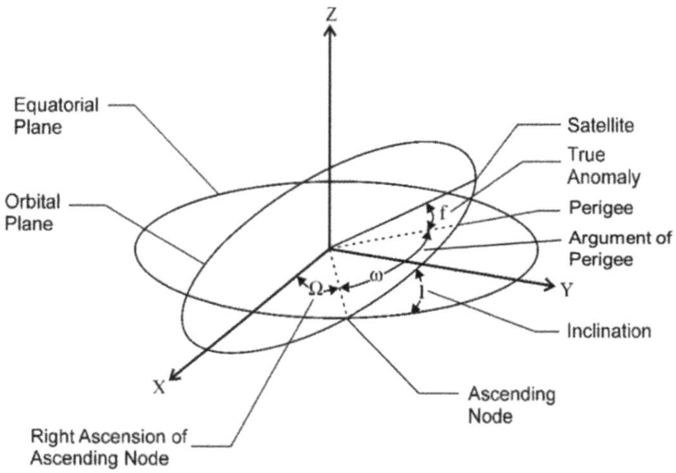

Figure 1.2.2.1 Orbital elements of a satellite in orbit around the earth.

For a given inclination, there can be an infinite number of orbital planes. The intersection of the equatorial plane and the orbital plane is known as the line of nodes. The right ascension of the ascending node along with the inclination uniquely specifies the orbital plane. The ascending node is the longitude at which the satellite crosses the equator while going from south to north. The descending node is where the satellite crosses the equator while going from north to south. Since the earth is spinning, we cannot use the common latitude/longitude coordinate system to specify where the line of nodes points. Instead, we use an astronomical coordinate system, known as the right ascension/declination coordinate system, which does not spin with the earth. Right ascension is an angle measured in the equatorial plane from a reference point in the sky where right ascension is defined to be zero, which is the vernal equinox. Thus the right ascension of the ascending node Ω denotes an angle, measured at the centre of the earth, from the vernal equinox to the ascending node.

The true anomaly f is the angle at the centre of the earth, measured in the orbital plane, between the perigee and the satellite at any given time. If the satellite is in a circular orbit and hence moving at a constant speed, this angle would point directly to the satellite.

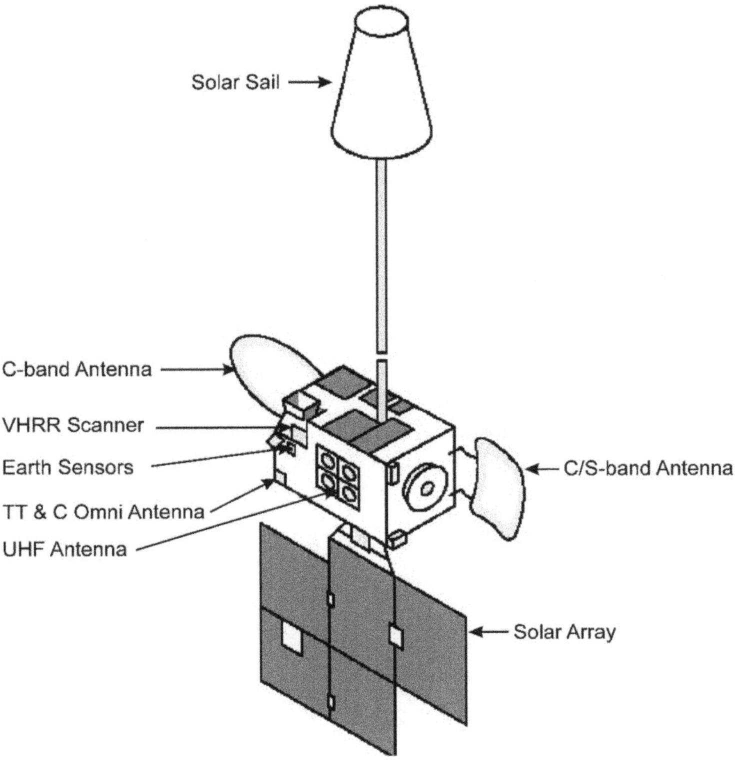

Figure 1.2.2.2 A schematic of the INSAT-1 spacecraft. (Source: IMD)

A satellite orbit may be subjected to small perturbations due to different reasons such as non-sphericity of the earth's geoid leading to gravitational anomalies, gravitation effects of other celestial bodies like the sun, moon or stars, atmospheric drag, solar radiation pressure and tides. The order of magnitude of these perturbations would depend on the nature of the satellite orbit. For example, geostationary satellites at 36,000 km height would be less prone to atmospheric drag, but more sensitive to the pressure of solar radiation, whereas the reverse would be true in the case of low altitude polar orbiting satellites. Early spacecrafts of the INSAT series had large solar panels on one side and a long solar sail on the other side for balancing the solar radiation pressure (Figure 1.2.2.2).

1.2.3 Types of Orbits

For almost four decades, from TIROS-1 in 1960 to the Tropical Rainfall Measuring Mission (TRMM) satellite in 1997, there were only two preferred orbits for meteorological satellites, polar and geostationary. A polar orbit is typically a circular orbit having an altitude of about 850 km and a period of revolution of about 100 min (Figure 1.2.3.1). This results in the satellite scanning a 3000 km wide swath on the earth's surface. It crosses the equator only twice a day, but views the poles in every orbital revolution. With each successive revolution, a new region of the earth comes under its view because in the meantime the earth has rotated around its own axis. The satellite completes 14 orbits in a day, so a global montage of images can be constructed over a period of 12 or 24 hr.

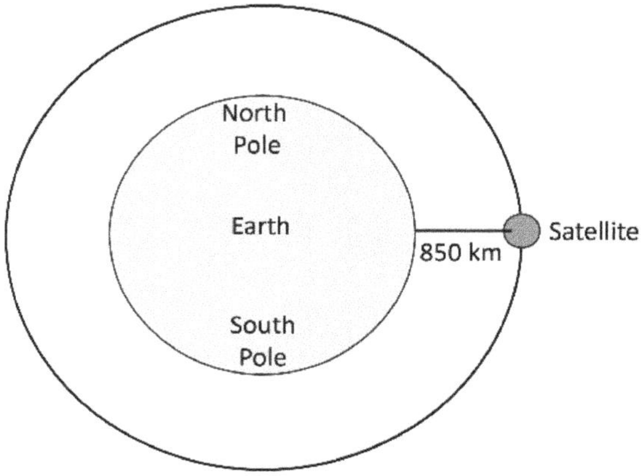

Figure 1.2.3.1 Polar orbit.

The earth is not a perfect sphere and the surface mass is not uniformly distributed all over, leading to gravitational anomalies. Other bodies in the solar system also exert their gravitational influence on a spacecraft in orbit around the earth. It is possible, however, to use these factors to advantage and induce precession or a slow movement of the satellite orbital plane with respect to the fixed inertial space. A special type of a polar orbiting satellite is a sun-synchronous satellite which is designed in such a manner that the precession of the satellite orbit matches the earth's revolution around the sun. The result is that the satellite crosses a place on the equator at the same local time (there are AM and PM orbits) and views it under comparable lighting conditions. The orbit does not go exactly over the poles but needs to have an inclination of about 99 or 100°. Occasional orbit manoeuvres may be required to maintain the sun-synchronous characteristics of the orbit.

When a satellite is placed in a circular orbit with a radius of 42,400 km or a height of 36,000 km above the earth's surface, it will circle round the earth with the same angular velocity as that of the earth. At that specific height, it will complete one orbital revolution around the earth in 24 hours. Such a satellite is called a geosynchronous satellite (Figure 1.2.3.2). If its orbit lies in the earth's equatorial planet it will appear, in a relative sense, to remain stationary above a given point on the equator. Such a satellite is called a geostationary satellite and it can provide a continuous earth view. Practically, however, the useful view is limited to about 60° around the subsatellite point, as the image outside it gets distorted. A constellation of 5 or 6 geostationary satellites spaced around the equator can together give a near-global picture. Geostationary satellites are ideal for communications purposes and continuous monitoring of the weather over the region within their view. Considerations of orbit stability demand that a geostationary satellite be placed only over a point on the equator and not above any other latitude. To maintain the geostationary satellite within the nominal limits of altitude and inclination, station-keeping manoeuvres have to be carried out periodically.

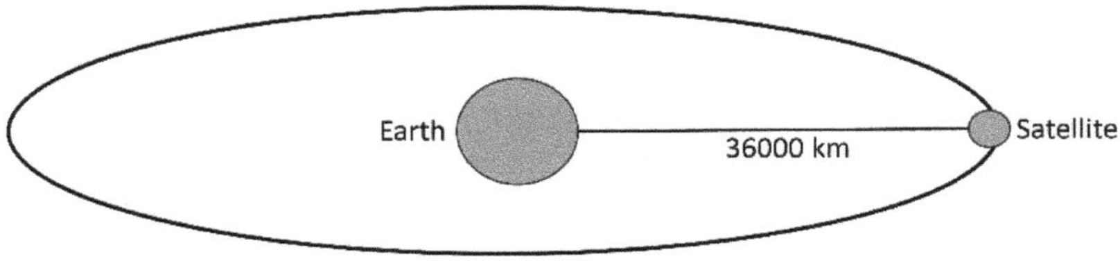

Figure 1.2.3.2 Geostationary orbit.

When the direction in which a satellite is moving in its orbit around the earth is the same as that of the rotation of the earth, the orbit is called direct or prograde. If the satellite is moving in a direction that is opposite to that of the earth's rotation, the orbit is called retrograde. A geostationary satellite orbit is generally prograde and a polar orbit generally retrograde.

Kelkar et al (1982) had envisaged a low-altitude non-geostationary satellite that would orbit around the equator and provide frequent meteorological coverage of the entire tropical belt. In the years that followed, many such orbits with different inclinations have become a reality. In 1997, TRMM was launched into an orbit having 35° inclination and in 2011, Megha-Tropiques in a 20° inclined orbit. In 2014, the GPM core satellite went into a 65° inclination orbit and in 2023, TROPICS into a 30° inclination orbit. These missions are described in later chapters. The orbital inclinations of these satellites are shown schematically in Figure 1.2.3.3.

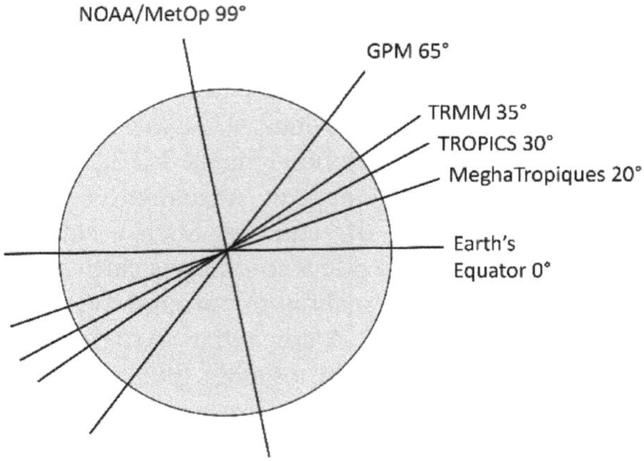

Figure 1.2.3.3 Orbital inclinations of various satellites.

While geostationary meteorological satellites provide continuous coverage of the earth, they have many basic constraints that simply cannot be overcome. For a geostationary satellite to be stable, its orbit must ideally lie in the earth's equatorial plane. Its subsatellite point can only be on the equator and over no other latitude. So one cannot have a satellite positioned over say, New Delhi or New York. With more countries launching their own geostationary satellites, the already crowded equatorial parking space is also going to come under greater stress.

A ingeniously designed orbit that combines the advantages of both polar and geostationary orbits is known as the Molniya orbit and between 1965 and 2004, several Russian satellites were launched into it. An ideal geostationary orbit must be perfectly circular, that is its eccentricity should be zero. In contrast, a Molniya orbit is by definition highly eccentric. Its perigee is about 400 km like that of a low earth orbit satellite and its apogee is of the order of 40,000 km like a geostationary satellite (Figure 1.2.3.4). Its orbital period is about 12 hr. In accordance with Kepler's second law, a spacecraft in a Molniya orbit would remain for a long time in the vicinity of the apogee. If the apogee is above a polar region, then a long continuous view of the polar region can be obtained.

The Molniya satellites were used mainly for purposes like communications and their application to meteorology remained limited. Another constraint was that spacecrafts in a Molniya orbit had to pass through the earth's Van Allen radiation belts repeatedly and therefore needed appropriate radiation shielding. Currently two Russian Molniya satellites Arktika-M-N1 and -N2, launched in 2021 and 2023 respectively, are operational and a new Molniya satellite series named Meridian has also been planned.

The advantages of having two or three satellites in highly elliptic orbits for observing the polar regions have been discussed in detail by Trishchenko et al (2019).

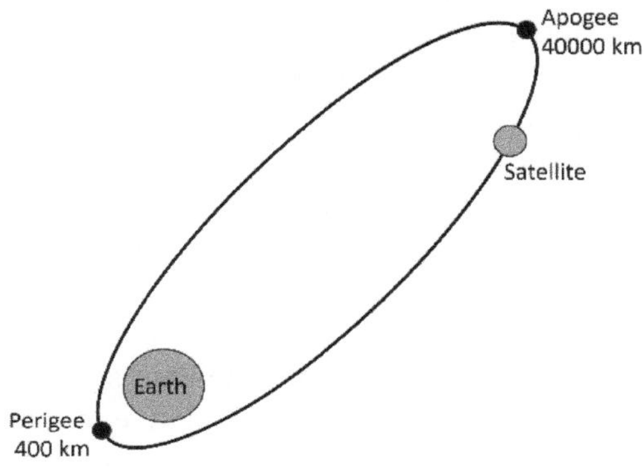

Figure 1.2.3.4 Highly elliptical Molniya orbit.

Around the beginning of the new millennium, the World Meteorological Organization (WMO) had mooted the concept of an International Geostationary Laboratory (IGeoLab). This called for a partnership among various countries that would allow geostationary meteorological satellites to be moved around to different locations on the equator as required, either for demonstration missions or for operational reasons (Hinsman 2006). A major logistical challenge to the exercise, however, was the non-availability of a sufficient number of the equatorial slots. While there was a strong interest in the meteorological satellite community, the IGeoLab project did not eventually materialize for various reasons. One of the missions proposed under the IGeoLab programme, was the Geostationary Imaging Fourier Transform Spectrometer (GIFTS). This was a new measurement concept that combined several advanced technologies, providing high vertical resolution temperature and moisture soundings (Smith et al 2002). This too showed much promise but did not make enough headway.

1.2.4 Satellite Attitude

While the altitude of a satellite is its height above the earth, the term satellite attitude refers to its orientation in space. Altitude and attitude are two equally important characteristics of a satellite and they together determine the kind of mission that the satellite can best perform. In the case of a meteorological or remote sensing satellite, the antennas must always remain in ground contact for communications and data transfer, and its imaging sensors must always view the earth as intended. Any deviation from the nominal attitude would result in an improper functioning of the satellite. Satellite attitude therefore demands to be carefully controlled and the two ways of doing it are through spin-stabilization and three-axes stabilization.

A spin-stabilized satellite rotates around its own vertical axis, spinning like a top and resisting external perturbation forces. Here the orientation of the satellite spin vector in space defines its attitude. Spin-stabilized satellites are equipped with thrusters which can be fired occasionally to bring about desired changes in the spin rate and to restore the spin vector orientation. The early geostationary satellites had a cylindrical design and rotated at the rate of one revolution per second during which the earth's disc would be scanned. A disadvantage in spinning satellites is that they cannot have large solar arrays and require to be supported by battery power. Another inconvenient factor is that instruments or antennas are required to de-spin so that the antennas and radiometers maintain their desired orientation relative to the earth.

The first Indian National Satellite, INSAT-1A, launched in 1982, was different from the U. S. and European geostationary satellites operating at that time, in that it was a non-spinning satellite, a design that was later adopted by many other countries including the U. S. In a non-spinning satellite, the attitude is defined in terms of the deviations from the nominal orientations of its roll, pitch and yaw axes. The terms roll, pitch and yaw are applied to a satellite in the same manner as they are used in the case of sailing ships or aircrafts in flight. Roll is the rotation of the spacecraft around the direction of its forward movement. Yaw is its rotation around the axis looking at the nadir. Pitch is the rotation around an axis normal to the roll and yaw axes. The three axes can be envisaged as lines running through the satellite's centre of gravity and intersecting at right angles (Figure 1.2.4.1).

The attitude of a non-spinning satellite is controlled by minimizing the roll, pitch and yaw. This is called three-axes stabilization, and achieved through the deployment of electrically powered spinning wheels called momentum or reaction wheels. These wheels are mounted in three orthogonal axes on the spacecraft and allow a transfer of angular momentum back and forth between the spacecraft and wheels. If the satellite is found to be deviating from its desired attitude, the appropriate spinning wheels are speeded up or slowed down to restore the correct attitude. Spacecrafts may also have propulsion system thrusters to apply the required torque. An advantage of three-axes stabilization is that radiometers and antennas can always be made to point at the desired targets without having to perform de-spin manoeuvres.

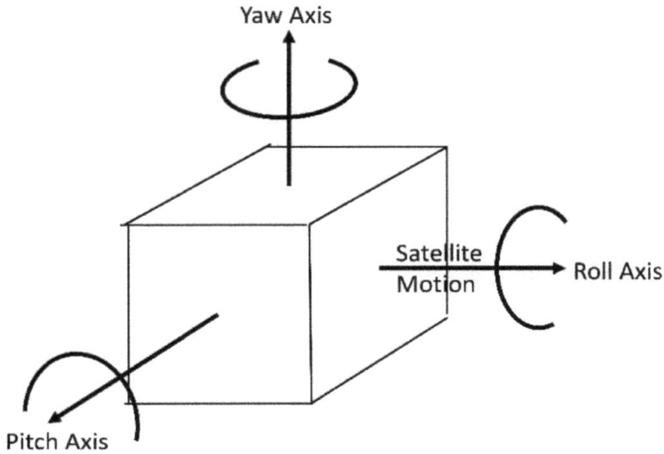

Figure 1.2.4.1 Roll, pitch and yaw axes of a satellite.

In three-axes stabilized geostationary meteorological satellites, the scanning radiometers have to alternate their scans between east-west and west-east directions until the earth's disc gets fully scanned. The polar-orbiting NOAA satellites are also three-axes stabilized but the scanning is required to be done only in the east-west direction as the north-south progression of the scan is taken care of by the movement of the satellite in its orbit.

Attitude errors result in a distortion of the satellite image and the scene viewed looks slightly different from the nominal view. Yaw errors will result in the image getting rotated around the sub-satellite point in a clockwise or anticlockwise manner. With roll and pitch errors, the image will not be centred at the nominal sub-satellite point but a little away from it. When the attitude errors are large, the sub-satellite point of a geostationary satellite is seen to meander in a figure-of-eight pattern around its nominal position on the equator.

1.3 Meteorological Satellite Payloads

The basic payload carried by the first ever meteorological satellite in 1960 was just a television camera that relayed to the ground whatever it saw of land, ocean and clouds. With every successive meteorological satellite launched over the last six decades by different countries, there have been rapid advances in satellite instrumentation and technology, serving various areas of application, which will be described in the next few sections and the following chapters of this book.

1.3.1 TV Cameras

Historically speaking, the earliest attempt to observe the earth's weather from space was made by the U.S. satellite, Vanguard 2, which was launched on 17 February 1959 but it could not collect much useful data. The world's first meteorological satellite to be considered a success, was the Television and Infra-Red Observation Satellite, TIROS-1, which was launched by the U. S. on 1 April 1960. TIROS-1 was operational for only 78 days, but it demonstrated beyond doubt the tremendous potential use of satellites for surveying global weather conditions from space and paved the way for future advancements.

TIROS-1 had carried two television cameras, a low-resolution one and a high-resolution one. The pictures were transmitted directly to ground receiving stations. Each camera had a magnetic tape recorder for storing photographs taken when the satellite was out of range of ground stations. The spacecraft was spin-stabilized but not earth-oriented. Therefore, the cameras were only operated while they were pointing at the earth when that portion of the earth was in sunlight. The video systems relayed thousands of pictures containing cloud cover views of the earth that provided information about large scale cloud regimes for the first time.

Many more satellites in the TIROS series were launched thereafter. TIROS-8 launched in December 1963 was the first satellite to be equipped with the Automatic Picture Transmission (APT) capability that made satellite pictures available the world over to meteorological agencies in real time as the satellite passed over their region. For this purpose, 50 ground stations were established, including one at Mumbai (Colaba) in India (Datar et al 1971),

The successor satellite programme was named after the U.S. agency, the Environmental Science Services Administration (ESSA), which operated it. The ESSA series was initiated in 1966 as an extension of the TIROS programme and its primary objective was to acquire higher resolution cloud cover images. Over the next four years, nine satellites in the ESSA series transmitted thousands of images that helped to predict weather patterns including hurricanes. By the end of the programme, more than 300 APT reception stations had come up in 45 countries.

In 1964, the U.S. National Aeronautics and Space Administration (NASA) launched a parallel satellite programme named Nimbus, to develop observation systems aimed at meeting the research requirements of atmospheric and earth sciences. Between 1964 and 1978, NASA launched seven Nimbus satellites which had very advanced on-board instruments for mapping ozone, sea ice, radiation budget components, coastal zone properties and sea surface temperature, However, Nimbus satellites also continued to carry an improved vidicon camera system for daylight coverage of local cloud conditions which were transmitted through the APT system.

A similar APT service was provided parallelly by the Russian Meteor series of satellites. However, the early satellites in this series were not sun-synchronous and the images were received at different times of the day in each pass.

1.3.2 Scanning Radiometers

After the end of the TIROS satellite series, TV cameras were given up and replaced by scanning radiometers which have since became more and more advanced. Scanning radiometers do not take snapshot pictures of the earth's surface and clouds but they scan them and measure the radiation received from them. In spin stabilized satellites, the rotation of the satellite about its spin axis provides the east-west sweep of the detectors. In 3-axes stabilized satellites, the detectors have to alternately perform east-west and west-east scans. The north-south scan is achieved in mechanical steps.

The digital data so collected is processed on ground and converted into an image format. What the resultant satellite image looks like depends upon the spectral band in which the satellite radiometer has measured the radiation, the sensitivity and response of the sensor, the height from which the satellite is viewing the earth, the time of the scan and such other factors.

The TIROS-N satellite launched by the U.S. in October 1978, carried a 4-channel radiometer known as the Advanced Very High Resolution Radiometer (AVHRR). This was subsequently improved to a 5-channel instrument (AVHRR/2) that was initially carried on the NOAA-7 satellite launched in June 1981. The AVHRR/3 with 6 channels was first carried on NOAA-15 in 1998. Details of the 6-channel NOAA AVHRR/3 are given in Table 1.3.2. The spectral bands are window regions of the spectrum and are so chosen as to allow maximum amount of radiation to reach the satellite. Most subsequent polar orbiting and geostationary satellites have continued with a similar or expanded version of this basic channel configuration in their scanning radiometers.

Details of the Very High Resolution Radiometer (VHRR) instrument on successive INSAT satellites are summarized in Table 1.6.1. The design and functioning of the INSAT INSAT-2A and -2B VHRR instrument have been described by Joseph et al (1994), further improvements and additions made to the INSAT-2E VHRR by Iyengar et al (1999) and the far more sophisticated INSAT-3D imager operations by Mohapatra et al (2021).

Table 1.3.2.1 NOAA AVHRR/3 Channels.

Channel Number	Spectral Band	Wavelength Range (μ)	Applications
1	VIS	0.58 - 0.68	Daytime cloud and surface mapping
2	NIR	0.725 - 1.00	Land-water boundary delineation
3A	NIR	1.58 - 1.64	Snow and ice detection
3B	MIR	3.55 - 3.93	Night-time cloud mapping, sea surface temperature retrieval
4	TIR	10.3 - 11.3	Night-time cloud mapping, sea surface temperature retrieval
5	TIR	11.5 - 12.5	Sea surface temperature retrieval

1.3.3 Sounders

The most widely used instrument for measuring pressure, temperature and humidity at various levels in the atmosphere is the radiosonde. This is an electronic package of sensors and a transmitter that is attached to a balloon. As the balloon rises, the signals received at a ground station are decoded to obtain the vertical profiles of temperature, humidity and other parameters. Most upper air stations take two radiosonde ascents per day at the synoptic hours of 00 and 12 UTC. With the advent of satellites, sounders were designed to do what radiosondes did. Satellite-based soundings had several advantages like they did not need consumables such as balloons, gas and electronics, and they could be done frequently and at more grid points.

In 1979, the NOAA-6 satellite first carried an instrument called the TIROS Operational Vertical Sounder (TOVS) which in fact consisted of three separate parts: High Resolution Infra-red Radiation Sounder Version 2 (HIRS/2), Microwave Sounding Unit (MSU) and Stratospheric Sounding Unit (SSU). The system as a whole provided atmospheric temperature profiles from the surface to 10 hPa, water vapour content at three levels of the atmosphere and total ozone content. The HIRS/2 was a step-scanner 20-channel spectrometer with 19 IR channels and one VIS channel (Smith et al 1979). Sounders came into wide use subsequently in several orbiting and geostationary satellites.

In principle, a satellite sounder instrument makes radiance measurements in the absorption bands of CO_2 centred at 15 and 4.3 μ, water vapour at 6.7 μ and ozone at 9.6 μ. Sampling at the centre of the band yields radiation from the upper levels of the atmosphere, as radiation from below has already been absorbed by the atmospheric gas. Sampling away from the centre of the absorption band yields radiation from successively lower levels of the atmosphere. Details of the retrieval process will be described later in the next chapter.

The current INSAT-3D satellites are also equipped with a 19-channel sounder. The sounder has 18 narrow spectral channels in shortwave infrared, middle infrared and thermal infrared regions and one channel in the visible region. The ground resolution at nadir is nominally 10 km for all nineteen channels. The instrument can make soundings at 10 km ground resolution every 3 hours for a full frame scan. Derivation of vertical profiles of temperature and humidity over 30 x 30 km areas is possible. Besides these vertical profiles, there are many other parameters than can be derived such as atmospheric stability indices, total precipitable water, and total column ozone.

1.3.4 CCD Imagers

While Charge Coupled Devices (CCDs) were widely used much earlier in remote sensing satellites for land resources applications, India was the first country to fly a 3-band CCD-based imager in geostationary orbit on its INSAT-2E satellite to complement the VHRR. A similar instrument was also flown on the INSAT-3A satellite. It provided co-registered images of the earth in VIS (0.62-0.68 μ), NIR (0.77-0.86 μ) and SWIR (1.55-1.69 μ) regions of the spectrum. The ground resolution of these images at the sub-satellite point was 1 km for all the three bands. The spectral bands as well as dynamic range and saturation radiance set points were so selected that the images could be used for meteorological applications as well as vegetation mapping and other earth resources applications.

The CCD imager was similar to the VHRR as far as the basic telescope design and scan mechanism were concerned. The separation of the three bands was achieved by two dichroic beam splitters. This optical configuration was designed for achieving identical footprints on the ground while accommodating two different sizes of detector elements. Flexible programmable scan modes allowed generation of images with up to 24 contiguous strips covering an area of 6300 $\times$ 6300 km on earth. Again, a positioning mechanism enabled this image field to be positioned anywhere in a scan field of 20° $\times$ 20° covering the full earth disc.

A complete technical description of the INSAT CCD imager has been given by Iyengar et al (1999), while Padmanabhan et al (2004) have discussed the complex nature of the CCD image processing. Bhatia et al (1999) have described the wide variety of applications possible with the fine resolution and continuous temporal coverage provided by the INSAT CCD imager.

1.3.5 Multispectral and Hyperspectral Imaging

The spectral resolution of a satellite sensor is inversely related to the spectral range covered by it. A higher spectral resolution and an accurate spectral signature can be obtained by having a large number of narrow bands, rather than one continuous broad band. Further, the spectral resolution and spatial resolution are also inversely related, since for registering the small amount of radiance contained in a very narrow spectral band, the sensor must dwell on the ground area for a longer time. However, if the satellite sensors are made to scan the earth at a slow speed they will take longer to complete the scan, thus reducing the temporal resolution or repetivity of observations as well. Remote sensing satellites can have a long repeat cycle of 10-20 days and can therefore make measurements in narrow spectral channels. Meteorological satellites cannot do this and in order to accomplish fast earth scans, they acquire stronger radiances from wider spectral bands such as VIS 0.55-0.75 μ or TIR 10.5-12.5 μ. The sensor response is, however, not uniform over the entire width of the channel and even if many satellite radiometers have nominally the same spectral bandwidths, their spectral response functions across the band are not identical. This is the reason why images of the same region obtained from different satellites look a little different from each other.

Precision spectroradiometers are usually based on a monochromator with a diffraction grating. A concave mirror collimates the radiation and the diffraction grating reflects the radiation while dispersing it into its spectral components. A second concave mirror focusses the radiation on a detector. Scanning of the spectrum is accomplished by rotating the grating while recording the electrical signal at the detector.

A satellite-based spectroradiometer can provide extremely high spectral sensitivity with exceptionally low out-of-band response. One such instrument called the Moderate Resolution Imaging Spectroradiometer (MODIS) was flown aboard the Terra and Aqua satellites launched in 1999 2002 respectively by NASA. Terra's orbit around the earth is timed so that it passes from north to south across the equator in the morning, while Aqua passes south to north over the equator in the afternoon.

Table 1.3.5.1 MODIS Spectral Bands and their Applications.

Band	Spectral Range (μ)	Primary Applications
1	0.620 – 0.670	Land/Cloud/Aerosols Boundaries
2	0.841 – 0.876	
3	0.459 – 0.479	Land/Cloud/Aerosols Properties
4	0.545 – 0.565	
5	1.230 – 1.250	
6	1.628 – 1.652	
7	2.105 – 2.155	

Table 1.3.5.1 *Contd....*

Band	Spectral Range (μ)	Primary Applications
8	0.405 – 0.420	Ocean Colour/ Phytoplankton/ Biogeochemistry
9	0.438 – 0.448	
10	0.483 – 0.493	
11	0.526 – 0.536	
12	0.546 – 0.556	
13	0.662 – 0.672	
14	0.673 – 0.683	
15	0.743 – 0.753	
16	0.862 – 0.877	
17	0.890 – 0.920	Atmospheric Water Vapour
18	0.931 – 0.941	
19	0.915 – 0.965	
20	3.660 - 3.840	Surface/Cloud Temperature
21	3.929 - 3.989	
22	3.929 - 3.989	
23	4.020 - 4.080	
24	4.433 - 4.498	Atmospheric Temperature
25	4.482 - 4.549	
26	1.360 - 1.390	Cirrus Clouds Water Vapour
27	6.535 - 6.895	
28	7.175 - 7.475	
29	8.400 - 8.700	Cloud Properties
30	9.580 - 9.880	Ozone
31	10.780 - 11.280	Surface/Cloud Temperature
32	11.770 - 12.270	
33	13.185 - 13.485	Cloud Top Altitude
34	13.485 - 13.785	
35	13.785 - 14.085	
36	14.085 - 14.385	

As of July 2025, both Terra and Aqua were still operational. MODIS views the entire earth's surface every 1 to 2 days, acquiring data in 36 spectral bands ranging in wavelength from 0.4 to 14.4 μ. The responses are tailored to meet the needs of several applications (Table 1.3.5). The resolution at nadir is 250 m for bands 1-2, 500 m for bands 3-7 and 1 km for bands 8-36. A ± 55° scanning pattern at an orbital altitude of 705 km achieves a 2,330 km swath.

Hyperspectral imaging is a powerful means of further reducing the spectral channels to extremely small widths, even as small as 0.01 μ. With such narrow widths, the spectral range from 0.38 to 2.55 μ for example, can be divided into as many as 217 intervals. This is made possible by the use of silicon microchips as detectors for the very near IR region, and Indium-Antimony (In-Sb) alloy detectors for the shortwave SWIR region. The radiance values measured by the CCDs for each such narrow interval are plotted as a graph of intensity versus wavelength, that provides a sufficient number of points through which we a meaningful continuous spectral curve can be drawn. Hyperspectral imaging is in principle far superior to multi-spectral imaging, as it provides much more detail about the spectral properties of the features to be identified. The hyperspectral data yields continuous spectral signatures rather than the band histogram plots.

1.3.6 Passive Microwave Sensors

Seasat, Nimbus-5, Nimbus-7 and India's Bhaskara-II, were the earliest satellites to carry passive microwave radiometers and clearly demonstrated the potential of microwave measurements in the retrieval of atmospheric and oceanographic parameters.

Bhaskara-II, which was launched in 1981, carried a Satellite Microwave Radiometer (SAMIR) that had three channels at 19.35, 22.235 and 31.4 GHz with vertical polarization. The observations were taken close to nadir at angles of 2.8° and 5.6° along the satellite ground trace during each spin of the satellite. The footprint had a size of 240 km. Pathak et al (1992) used SAMIR data for the period January 1982 to June 1983 to derive the latitudinal variation of monthly mean values of atmospheric water vapour content over the Arabian Sea and Bay of Bengal and studied its seasonal and geographical variations. Gohil et al (1982) had earlier derived atmospheric water content from Bhaskara SAMIR data.

Microwave remote sensing, however, took some time to come into wider use. This was because of the weak microwave signal strength, poor ground resolution of microwave images and the high variability of the land surface emissivity in the microwave region. The principles of satellite-borne microwave radiometry have been reviewed by Pandey (1995).

The Defense Meteorological Satellite Programme (DMSP) is a long term U.S. Air Force effort in space to monitor the meteorological, oceanographic and solar-geophysical environment of the earth. In December 1972, DMSP data was declassified and made available to the civilian and scientific community. The Special Sensor Microwave Imager (SSM/I), was first flown on the DMSP satellites in June 1987. This was a 7-channel, 4-frequency, linearly polarized, passive microwave radiometric system measuring atmospheric, ocean and terrain microwave brightness temperatures at 19.35, 22.235, 37.0 and 85.5 GHz with footprint sizes varying from 13 × 15 km at 85 GHz to 43 × 69 km at 19 GHz.

The TRMM Microwave Imager (TMI) on board the Tropical Rainfall Measuring Mission (TRMM) satellite, launched in 1997, was a passive sensor operating at 5 microwave frequencies of very similar to those of SSM/I. The Indian Remote Sensing Satellite IRS-P4, also known as Oceansat-1, which was launched in May 1999, carried a Multi-channel Scanning Microwave Radiometer (MSMR) operating at 6.6, 10, 18 and 21 GHz frequencies in both H and V polarizations.

Microwave data are now being used to derive a variety of geophysical parameters such as, ocean surface wind speed, ice cover, cloud liquid water, integrated water vapor, precipitation over water, soil moisture, land surface temperature, snow cover and sea surface temperature. Most of the retrieval algorithms in vogue are in the form of statistical correlations between the brightness temperatures of various channels or differences between channels, with these parameters.

1.3.7 Active Sensors

Passive scanning radiometers measure the radiation reaching the satellite from the earth's surface or cloud tops in different wavelength bands including microwave. In active remote sensing, on the other hand, a radar placed on board a satellite emits microwave signals at a certain frequency and then measures the radiation scattered back from the earth's surface or cloud tops. This yields much more information than a passive microwave radiometer would.

However, there are practical issues. For a radar on a geostationary satellite, the range becomes enormous. Polar orbiting satellites move at high speed relative to the earth and the dwell time is insufficient. Therefore, the beam width of the radar has to be narrow enough to be fully filled with precipitation and the swath has to be wide enough to eliminate gaps in coverage. A space-borne radar must also be able to discriminate the weak precipitation echo over the strong return from the earth's surface. Hence it becomes desirable to use wavelengths shorter than 3-10 cm which are used in conventional weather radars on the ground, as at shorter wavelengths the return echo is stronger.

The TRMM Precipitation Radar was the first space-borne radar that was designed specifically to measure rainfall in three dimensions over tropical and subtropical region.

Satellite measurement of cloud properties became a reality with the successful launch of two satellites, CloudSat and CALIPSO, in April 2006. CloudSat carried a mm-wave radar and CALIPSO had on board a backscattering lidar. The pair of satellites formed a unique combination of radar and lidar in space.

1.4 Satellite Imagery

As satellite images started becoming available to meteorologists around the world in the early 1960s, the skills of satellite image interpretation developed very rapidly. Until then, knowledge about clouds had been documented into cloud atlases on the basis of what had been observed from the ground or aircrafts. The view from satellite altitudes was, however, completely new, and clouds seen by satellites had to be interpreted in an altogether different manner.

In the early days of the satellite era, the imaging technology was primitive. APT pictures received directly from satellites flying overhead were produced on facsimile chart paper, mostly of inferior quality and liable to fading in a short time. The images many times lacked clarity and hence they had to be examined with great care. However, the technology improved really fast and even photographic paper was mostly done away with, when interactive computer image processing systems became common. Now of course, there are web sites on the internet and mobile apps on which anyone can access satellite imagery in near real time.

Satellite images are constructed by assigning the measured radiance a certain brightness value usually on a scale ranging from 0 to 255 or from 0 to 1023, which is called the gray scale. The number 0 on the gray scale stands for pure black and the highest number stands for pure white.

In a typical VIS image (Figure 1.4.1), the lowest values on the gray scale are seen over oceans because of their high absorption of solar radiation. The highest gray scale values are seen over snow covered regions and Cb cloud tops, which reflect solar radiation the most. Shallow clouds having low cloud water/ice content and large cloud droplet size have intermediate gray scale values. In an IR image (Figure 1.4.2), the lower end of the gray scale corresponds to high temperatures and the higher end of the gray scale to low temperatures. In a WV image (Figure 1.4.3), lower values of the gray scale correspond to dry regions and higher values to moist regions.

The information derived from the VIS, IR, WV and microwave channels is not redundant but complementary. The most important advantage of IR and WV imagery is that it is available at all times unlike VIS imagery which is available only in daylight hours.

Figure 1.4.1 INSAT-3DS 3 August 2025 0500 UTC VIS full disc image. (Source: https://mausam.imd.gov.in/)

Figure 1.4.2 INSAT-3DS 3 August 2025 0500 UTC TIR1 full disc image.
(Source: https://mausam.imd.gov.in/)

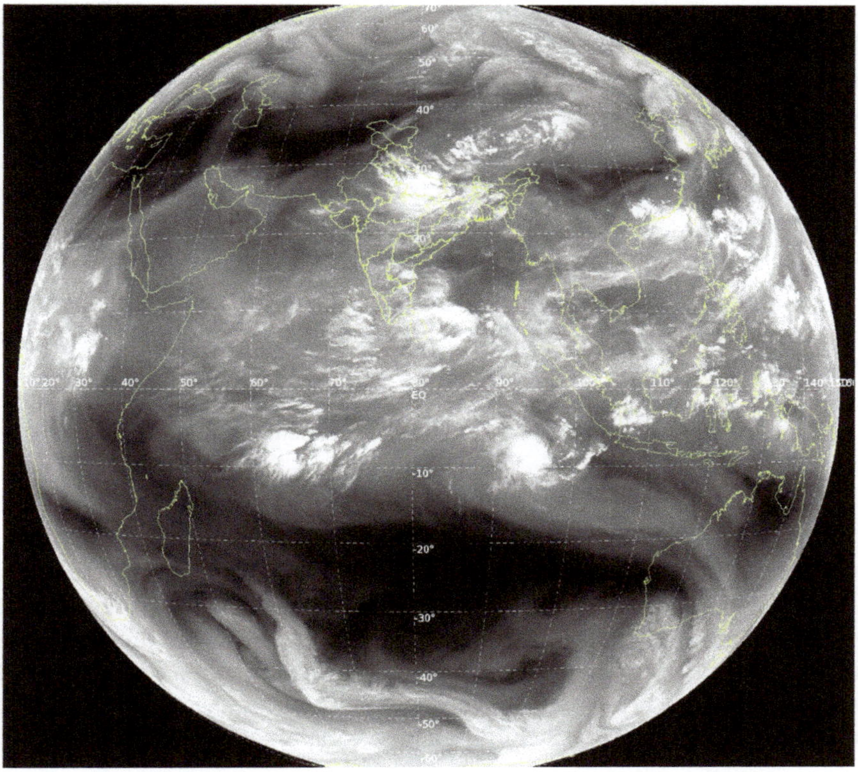

Figure 1.4.3 INSAT-3DS 3 August 2025 0500 UTC WV full disc image. (Source: https://mausam.imd.gov.in/)

1.4.1 Resolution

The term resolution as applied to meteorological satellites assumes a different meaning in different contexts: spatial, spectral, radiometric and temporal, but in each case it is a critical design consideration. While the design aim would be to have the finest resolution, it is not always possible to attain it and what is realized is a trade-off between several competing factors. A higher resolution also requires extremely high rates of data transfer and an increasingly complex design of both space and ground systems.

What a scanning radiometer views at any instant is called the Instantaneous Field Of View (IFOV) and it is generally considered as the spatial resolution of the imaging sensor. The size of an individual pixel or picture element also depends mainly on the sampling rate and the forward motion of the spacecraft. In a meteorological satellite, the spatial resolution of the radiometer will decide the smallest cloud that can be seen in the image.

The spectral resolution of the sensor is inversely related to the channel bandwidth. A higher spectral resolution and an accurate spectral signature can be obtained by having a large number of narrow bands, rather than one continuous broad band.

The term radiometric resolution is used to indicate the smallest perceptible change in the radiance of various targets that the sensor can discriminate. It is a measure of the signal-to-noise ratio of the sensor. It can also be defined in terms of the ability to resolve temperature difference between targets.

The temporal resolution is an important consideration in operational meteorology because it determines what kind of meteorological phenomena can be viewed by the satellite, as they have different lifetimes. Temporal resolution refers to the elapsed time interval between two successive satellite scans of a given region. This is also called the repeat cycle or repetivity of the orbit.

The temporal resolution is decided by the design of the scanning radiometer as well as the nature of the spacecraft orbit and it has to have a significant trade-off with the spatial resolution. For achieving a finer spatial resolution, the scan has to be made slower and the temporal resolution will get reduced. In a rapid scan, a selected area can be scanned repeatedly, but then other areas will be left out. Satellites meant for monitoring of earth resources or for cartographic applications can have an extremely high spatial and spectral resolution because they can afford to have a long repeat cycle.

1.4.2 Navigation and Gridding

In some VIS and IR satellite images, geographical features like coastlines, islands and lakes are distinctly seen, but many times they may be obscured because of clouding or a lack of temperature contrast. In WV images they are rarely noticeable. Therefore, for any satellite image to be used for practical purposes, it must be provided with a geographical reference frame. This last stage of the process of satellite image production actually involves three steps, navigation, gridding and registration.

The term navigation, also called geolocation, is used in its conventional sense as in flying, shipping or driving on a road. Image navigation makes it possible to relate every pixel, identified by its line and element number to a point on the earth, defined by its latitude and longitude, and vice versa.

By gridding is meant the superposition of names of locations, latitude-longitude lines or political boundaries on the navigated image. Once the mages are accurately geolocated, they can be transformed into desired cartographic projections like Mercator or Lambert and overlaid with maps.

If geostationary satellite images are viewed in quick succession, the surface features should appear to be stationary. The removal of apparent earth motion from an animated sequence of images is called registration.

The navigation and gridding process for a 3-axes stabilized geostationary satellite like INSAT has been explained in detail by Kelkar et al (1980). Briefly, it involves transformations among the following five coordinate systems:

(a) inertial system, having its origin at the centre of the dynamical earth, and the x-axis pointing to the vernal equinox and z-axis normal to the equatorial plane,

(b) rotating system, similar to the inertial system but with the z-axis passing through the Greenwich meridian,

(c) local vertical system, having its origin at the centre of the satellite,

(d) body-centred system, defined in terms of the pitch, roll and yaw angles, and

(e) picture frame system, defined in terms of line and element number.

In the actual process, Ground Control Points (GCPs) or precisely located earth features are used, and Global Positioning System (GPS) information for pinpointing the satellite location.

The accuracy of navigation has a very important bearing in quantitative product derivation like cloud motion winds or in lightning detection, in which the positions of the pixels have to be known very precisely. A poor navigation shows up as a mismatch between the superimposed grid and the geographical features on the image. Navigation errors may also result in image distortions. Channel-to-channel registration, frame-to-frame and within-frame registration are therefore all important (Tan et al 2020).

1.5 Indian Meteorological Satellites

The need for India to have its own meteorological and remote sensing satellite programme to monitor its weather and vast natural resources was realized way back in the 1960s by great pioneers and visionaries like Vikram Sarabhai and Satish Dhawan. The successful launches of the first Indian satellites Aryabhata, Bhaskara-I, Bhaskara-II and APPLE between 1975 and 1981 were followed by the launch of INSAT-1A in 1982 and INSAT-1B in 1983. The first Indian remote sensing satellite IRS-1A was launched in 1988. The evolution of the meteorological component of the Indian space programme has been reviewed from time to time by Kelkar (1994a, 1994b, 2009, 2019) and Bhatia (2025).

INSAT is India's unique multi-purpose geostationary satellite system which was conceptualized to bring in satellite-based services in as many fields as possible in the shortest time. In contrast with other countries that had dedicated satellites for different purposes, INSAT ushered in a new era in communications, television and radio broadcasting, and meteorology with a single satellite. Since 1965, Indian weather forecasters were having a limited access to low resolution satellite imagery of foreign satellites through the APT facility. It was only with the launch of the INSAT-1A satellite in 1982 that satellite meteorology may be said to have truly come of age in India.

Table 1.5 gives the chronology of the evolution of the INSAT meteorological programme. While the INSAT-1 series consisted of multi-purpose satellites, some of the later satellites did not have any meteorological component. On the other hand, Metsat, launched in 2002 was exclusively dedicated to meteorological imaging. Likewise, INSAT-3D, -3DR and -3DS, launched in recent years, are primarily meteorological satellites without any communication transponders except for data relay and satellite-aided search and rescue. INSAT-2E was the first geostationary meteorological satellite to have a CCD payload. INSAT-3D was the first Indian satellite with a sounder.

Satellites in the INSAT-1 series were designed by ISRO but built in the U.S. and launched from abroad. Satellites in the INSAT-2 series were built indigenously but launched from abroad. Metsat, was launched from Sriharikota by ISRO's Polar Satellite Launch Vehicle (PSLV). It was later renamed Kalpana-1 in memory of Dr Kalpana Chawla, the India-born American astronaut who perished in the U.S. Space Shuttle Columbia disaster on 1 February 2003. Satellites in the INSAT-3 series have been built by ISRO and launched with its own Geostationary Launch Vehicle (GSLV).

Table 1.5.1 Chronology of the Evolution of the INSAT Meteorological Programme.

<table>
<tr><th>Name of Satellite</th><th>Launch Date</th><th>Meteor-logical Payload</th><th>Channel</th><th>Spectral Range (μ)</th><th>Resolu-tion (km)</th></tr>
<tr><td>INSAT-1A</td><td>10 April 1982</td><td>VHRR (Very High Resolution Radiometer)</td><td rowspan="4">VIS
IR</td><td rowspan="4">0.55-0.75
10.5-12.5</td><td rowspan="4">2.75
11</td></tr>
<tr><td>INSAT-1B</td><td>30 August 1983</td><td>VHRR</td></tr>
<tr><td>INSAT-1C</td><td>21 July 1988</td><td>VHRR</td></tr>
<tr><td>INSAT-1D</td><td>12 June 1990</td><td>VHRR</td></tr>
</table>

Table 1.5.1 *Contd....*

Name of Satellite	Launch Date	Meteor-logical Payload	Channel	Spectral Range (μ)	Resolu-tion (km)
INSAT-2A	10 July 1992	VHRR	VIS IR	0.55-0.75 10.5-12.5	2 8
INSAT-2B	23 July 1993	VHRR			
INSAT-2E	3 April 1999	VHRR CCD Charge Coupled Device Camera	VIS IR WV VIS NIR SWIR	0.55-0.75 10.5-12.5 5.7-7.1 0.62-0.68 0.77-0.86 1.55-1.69	2 8 8 1 1 1
Metsat/ Kalpana-1	12 September 2002	VHRR	VIS IR WV	0.55-0.75 10.5-12.5 5.7-7.1	2 8 8
INSAT-3A	10 April 2003	VHRR CCD	VIS IR WV VIS NIR SWIR	0.55-0.75 10.5-12.5 5.7-7.1 0.62-0.68 0.77-0.86 1.55-1.69	2 8 8 1 1 1
INSAT-3D	26 July 2013	VHRR	VIS SWIR MWIR	0.55-0.75 1.55-1.70 3.80-4.00	1 1 4
NSAT-3DR	6 September 2015		WV TIR TIR	6.50-7.10 10.3-11.3 11.5-12.5	8 4 4
INSAT-3DS	13 February 2024	Sounder	SWIR MWIR LWIR VIS	3.67-4.59 6 channels 6.38-11.33 5 channels 11.66-14.85 7 channels 0.67-0.72 1 channel	10 10 10 10

1.6 References

Bhatia, R. C. and Gupta, H. V., 1999, "Use of charged coupled device payload on INSAT-2E for meteorological and agricultural applications", *Current. Sci.*, 1999, 76, 1444–1447.

Bhatia R. C. and Mitra A. K.. 2025, "Evolution of satellite meteorology and its future scope - India's perspective" *Mausam*, 76, 31-54.

Datar S. V. and Joseph C. P., 1971, "A satellite automatic picture transmission (APT) system ground receiving station", *Indian J. Meteor. Geophys.*, 22, 377-380.

Gohil B. S. and co-authors, 1982, "Remote sensing of atmospheric water content from Bhaskara SAMIR data", *Int. J. Remote Sensing*, 3, 235-241.

Hinsman D. E., 2006, "WMO Space Programme implementation activities", *14th Conf. Satellite Meteorology Oceanography*, Amer. Meteor. Soc.

Iyengar V. S. et al, 1999, "Meteorological imaging instruments on-board INSAT-2E", *Current Sci.,* 76, 1436-1443.

Kelkar R. R., Sant Prasad and Ellickson J., 1980, "Image navigation and gridding for three-axis stabilized geostationary satellites", *NOAA/NESDIS Report*, Washington DC, 32 pp.

Kelkar R. R., Sant Prasad and Khanna P. N., 1982, "Conception of an equatorial orbiting meteorological satellite for the tropics", *Mausam*, 33, 507-508.

Kelkar R. R., 1994a, "Satellites for monitoring climate change: the emerging scenario", *Proc. Indian National Science Academy*, 60, 335-348.

Kelkar R. R., 1994b, "Satellite meteorology in India – an overview", *Indian J. Radio Space Phys.*, 23, 235-245.

Kelkar R. R., 2009, "Evolution of satellite meteorology and its future scope", *Mausam, Diamond Jubilee Volume, Special Issue.*

Kelkar R. R., 2019, "Satellite meteorology in India: its beginning, growth and future", *Mausam*, 70, 1-14.

Mohapatra M. et al, 2021, "INSAT-3DR-rapid scan operations for weather monitoring over India", *Current Sci.*, 120, 1026-1034.

Padmanabhan N., Ramakrishnan R. and Gurjar S. B., 2004, "Geometric modelling of INSAT-2E CCD payload and multistrip mosaicing for geocoding large areas", *Current Sci.*, 86, 1113-1121.

Pandey P. C., 1995, "Satellite-borne microwave radiometry for atmospheric studies", *Indian J. Radio Space Phys.*, 24, 245-254.

Pathak P. N. and Gautam N., 1992, "Latitudinal distribution of water vapour over the Arabian Sea and Bay of Bengal using Bhaskara-II SAMIR data", *Mausam*, 43, 385-394.

Smith W. L. et al, 2002, "GIFTS-the precursor geostationary satellite component of the future Earth Observing System",. *Geoscience and Remote Sensing Symp. IGARSS'02*, 357-361.

Tan B. et al, 2020, "GOES-R series image navigation and registration performance assessment tool set", *J. Applied Remote Sensing*, 14, 032405-032405.

Trishchenko A. P., Garand L., and Trichtchenko L. D., 2019, "Observing polar regions from space: comparison between highly elliptical orbit and medium earth orbit constellations", *J. Atmos. Oceanic Techno.*, 36, 1605-1621.

Chapter 2

Weather Systems Observed in Satellite Imagery

The complexity of atmospheric motion arises from the rich variety and coexistence of atmospheric phenomena having space scales ranging from less than a metre to thousands of kilometres, and time scales shorter than a minute to longer than a week. They are manifested in various appearances, as tiny eddies and dust devils, short-lived thunderstorms and tornadoes, diurnal sea and land breezes, fronts, jet streams, tropical cyclones and long waves encircling the globe (Table 2.1). These varied atmospheric phenomena not only coexist, but also influence, modify and interact with those of smaller and larger scales. Our knowledge of the atmospheric phenomena is therefore limited by the number of times in a day that we take weather observations at a given place, and the distance between adjacent stations where these observations are made.

Operational global weather analysis and forecasting has conventionally been done on the synoptic scale, and over the years, the global observing network has evolved in such a manner as to meet the necessary and sufficient requirements of synoptic scale forecasting. Comparatively, observations on other scales, especially mesoscale observations, have received much lesser importance in the observational scheme. The global synoptic network has also remained essentially a land network supplemented by ships making some limited weather observations, because of the obvious difficulties in operating weather stations over the ocean that are difficult to overcome. It is only in recent times that ocean buoys and floats have begun to send observations for operational use.

Meteorological satellite imagery provides an additional means of monitoring the state of the atmosphere by visually examining clouds and cloud patterns. Images received from geostationary and orbiting satellites reveal directly or indirectly the presence of the wide spectrum of atmospheric phenomena just at a glance. The value addition of satellite imagery is the highest over the vast expanse of the world's oceans and inhospitable regions where conventional observations are not available.

2.1 Characteristics of Satellite Imagery

In the early years of satellite meteorology, six basic characteristics were used to identify clouds in satellite pictures (Conover 1962). These were:

(1) Cloud brightness relating particularly to the depth and composition,

(2) Texture - whether smooth, fibrous, opaque, or mottled,

(3) Form of elements - whether regular or irregular,

(4) Pattern of elements - associated with topography, air flow, vertical and horizontal wind shear,

(5) Size - both of the patterns and the individual elements, and

(6) Vertical structure - for example, shadows thrown below.

Later on, cloud classification schemes were developed on the basis of cloud patterns (Hopkins 1967):

(1) Vertical features:
   - (a) Circular or spiral bands,
   - (b) Crescent-shaped or comma-shaped cloud masses,
   - (c) Quasi-circular cloud masses, and
   - (d) Curved or linear bands.

(2) Major cloud bands, in which the length is much greater than the width,

(3) General features:
   - (a) Minor bands like cloud streets, jet stream bands, lee waves,
   - (b) Cumuliform features, polygonal cells, and
   - (c) Stratiform features, fog areas.

As of now most of the image interpretation work is carried out through digital processing, and various interpretation aids are available to the analysts. However, it is still helpful to have a basic knowledge of the fundamental image characteristics mentioned above.

Clouds which appear brighter in VIS imagery are those which have a large albedo because of their great vertical extent, high cloud water/ice content and small cloud droplet size. Shallow clouds having low cloud water/ice content and large cloud droplet size appear gray. Thus towering cumulonimbus clouds (Cb) appear the whitest in VIS images whereas thin cirrus can barely be seen (Figure 2.1.1).

The most important advantage of IR and WV imagery is that it is available at all times unlike VIS imagery which is available only in daylight hours. Secondly, IR radiances are a measure of the temperature of the radiating surface. Clouds which appear white in IR imagery are those which have cold tops such as Cb and cirrus clouds (Figure 2.1.2). A mature Cb cloud can be further recognized by its sharp edge on the windward side and a fuzzy edge on the other side resulting from the cirrus plume. The plume may get blown downwind over several hundred kilometres and give an indirect indication of the upper level wind speed and direction.

In WV imagery, the brighter regions are those with high upper tropospheric humidity and the dark regions are those where the upper troposphere is very dry. Generally speaking, WV images have a different appearance from that of VIS and IR images. WV images are characterized by the presence of extensive and continuous structures (Figure 2.1.3). Cloud patterns which appear distinct from each other in the VIS and IR images can be recognized as being a part of the same air mass in corresponding WV images. The band structures also provide some indication of prevailing large scale wind patterns like jet stream cores.

Land and ocean vary in their degree of grayness at different times and in different channels. In the VIS image of Figure 2.1.1 the ocean is dark because of its low albedo and land is gray. In the IR image of Figure 2.1.2 the hot desert land is black while the cooler ocean is gray.

It is also important to remember that interpretation of individual satellite images should never be done in isolation, as may happen for example, when one sees a satellite image flashed on a television channel or on the internet. The day and time of the image, the geographical area covered, the gray scale and enhancement used, must all be known and given due consideration. Other available observational data and synoptic weather charts should also always be referred to. Continuity in time is essential and earlier images should be compared for detecting changes or confirming the interpretation. Images from all available channels should also be compared.

The information derived from the VIS, IR, WV and microwave channels is not redundant but complementary, and when all channels are considered together, they help to remove uncertainties or ambiguities in the interpretation process and to identify clouds and surface features uniquely

The reflection of the sun's disc that may sometimes be seen in VIS channel satellite images of large water bodies is referred to as sunglint. It is particularly observed in noon time images of tropical oceanic regions taken by geostationary meteorological satellites. A smooth water surface acts like a mirror and produces a small bright reflection (Figure 2.1.4) while a rough water surface produces a weaker reflection over a larger area. Although scatterometers now give us direct measurement of the sea surface winds, in the very early years of satellite meteorology, attempts were made to estimate the sea surface wind from the area of sunglint and its brightness as seen in VIS satellite pictures (Gokhale 1971).

Figure 2.1.1 INSAT-3DS VIS image 22 July 2025 0600 UTC. (Source: https://mausam.imd.gov.in/)

Figure 2.1.2 INSAT-3DS TIR image 22 July 2025 0600 UTC. (Source: https://mausam.imd.gov.in/)

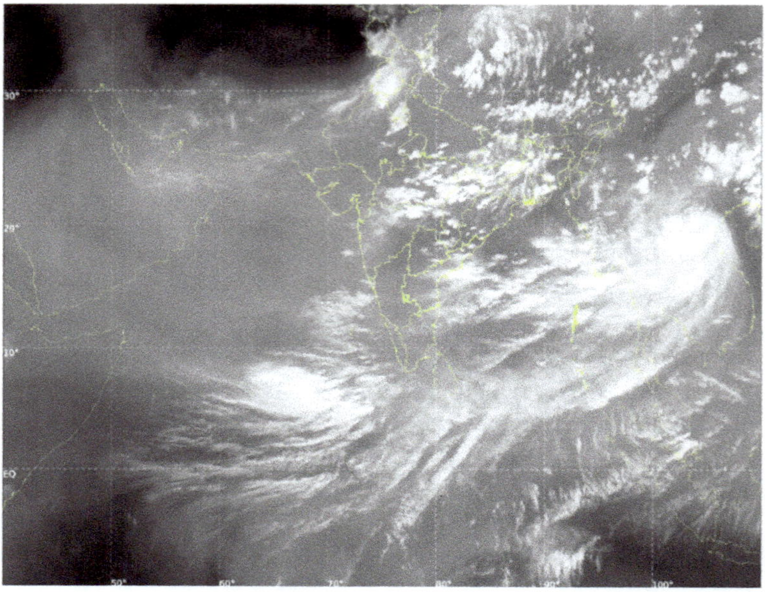

Figure 2.1.3 INSAT-3DS WV image 22 July 2025 0600 UTC. (Source: https://mausam.imd.gov.in/)

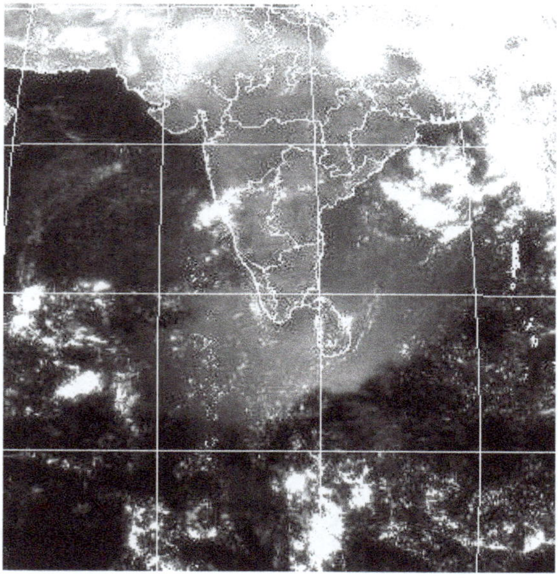

Figure 2.1.4 Kalpana-1 VIS image of 10 May 2006 showing a large area of sunglint over the ocean. (Source: IMD)

Table 2.1.1 Space and Time Scales of Atmospheric Phenomena.

Classification	Examples of Phenomenon	Approximate Space Scale	Approximate Time Scale
Global or Planetary Scale	Long waves, subtropical anticyclones, tropical easterly waves	5000-10000 km	Days to a week or longer
Synoptic Scale	Fronts, extratropical cyclones, anticyclones, cloud clusters, monsoon depressions, tropical cyclones	500-5000 km	Days to a week
Mesoscale	Lee waves, gravity waves, squall lines, land and sea breeze, mesoscale convective elements	20-500 km	Hours to days
Convective Scale or Small Scale	Thunderstorms, tornadoes, convective cells	1-20 km	Minutes to hours
Microscale	Boundary layer eddies, dust devils	100 m - 1 km	Seconds to minutes

2.2 Local Severe Storms

The space and time scales of atmospheric phenomena have been given in Table 2.1. It must be understood here that every scale of motion contains within itself smaller scales of motion and there is an interaction amongst them. To get a reasonable picture of a weather system, it is generally accepted that there should be at least 10 observations per wavelength. Present observational networks are designed for analysis of synoptic scale systems and not mesoscale systems for which special observational networks are required to be set up. Satellites are therefore most useful in observing and monitoring mesoscale systems that synoptic networks are unable to resolve. The several types of mesoscale convective systems like thunderstorms, hailstorms, dust-storms and tornadoes are discussed in subsequent sections.

2.2.1 Thunderstorms, Hailstorms and Cloudbursts

There are three basic conditions for the development of a thunderstorm: moisture, instability and a lifting mechanism. Air is considered unstable if it continues to rise when given an upward push. An unstable air mass is characterized by warm moist air near the surface and cold dry air aloft. As a rising parcel of air cools, some of the water vapour will condense, forming a cumulonimbus or Cb cloud that is commonly called the thunderstorm.

For a thunderstorm to develop, what is needed is an initial trigger or a mechanism that will give a start to the process of upward motion. The temperature of the lowest layers of the atmosphere increases rapidly in the afternoon or evening because of land heating and the warmest air tends to rise. Lifting is also provided by fronts, particularly cold fronts, and dry lines. Terrain may also be a cause for lifting, as when air currents encounter a mountain barrier, they are forced to rise up the slope of the terrain.

The basic building block of all thunderstorms is the thunderstorm cell. The thunderstorm cell has a distinct life cycle that lasts typically for about 30 min. The life cycle of a thunderstorm can be described a consisting of three stages:

(a) In the towering cumulus stage, a cumulus cloud begins to grow vertically, to a height of up to 6 km. Air within the cloud is dominated mostly by updrafts with some turbulent eddies around the edges.

(b) In the mature stage of the thunderstorm, the cloud can grow to a considerable height, often up to 12 km or above. Satellite pictures show occasional evidence of overshooting cloud tops which have even penetrated the tropopause. Rain is produced and the downward frictional drag force exerted by the falling rain drops on the surrounding air produces downdrafts. Cooling of the air due to evaporation of the rain drops further increases the speed of the downdrafts. Updrafts and downdrafts are both relatively weak, of the order of 10 m/sec, and they can interfere with each other.

(c) In the dissipating stage, rain spreads throughout the cloud and the downdraft becomes extensive. The updraft gets exhausted, the storm ceases to have the supply of warm moist air that is necessary to maintain itself, and the cloud dissipates. Light rain and weak outflow winds may remain for a while during this stage, before leaving behind just a remnant anvil top.

When the vertical wind shear is weak, single-cell thunderstorms occur. They last for about 30 min and get carried by the low level wind. When there is moderate wind shear, multi-cell storms develop. Here, the individual storms have a life of about 45 min, but a group of cells can exist for several hours with new cells forming and replacing old dissipating cells.

A severe thunderstorm has a life of a few hours and it produces very heavy rain, sometimes large hail, and strong squalls on the ground. The rain may be so intense as to cause flash floods. There is high vertical wind shear, both updrafts and downdrafts are very strong, up to 50 m/sec, and they co-exist without interference. Warm moist air enters the cloud from the front and rises. This results in strong updrafts which produce rain and support the formation of large hail. Dry air at the middle level enters the cloud from behind. Rain falls into it, and evaporation of rain cools the air, resulting in strong downdrafts.

The upper portion of the thunderstorm finally reaches such a high level that it is influenced by the strong winds in the upper troposphere. These carry the updrafts large distances away from the cloud, forming what is known as the thunderstorm anvil. A severe thunderstorm is 10-20 km wide and 12-18 km tall, while the anvil may extend horizontally up a distance of up to 100 km. Several such storms may get organized along a line called a squall line. Severe thunderstorms are of the multi-cell or super-cell type. In the super-cell type, the entire storm behaves as if it is a single cell. It may evolve into a rotation circulating, called a meso-cyclone, which can generate a tornado.

There are different motions associated with a thunderstorm: the updrafts and downdrafts within it, the motion of the thunderstorm cloud itself, the motion of the anvil relative to the thunderstorm, the squall speed and the motion of the gust front. Each of these has its own dynamics and scale. The speed and direction of motion of the thunderstorm are functions of the ambient wind in a deep layer. The speed and direction of movement of the anvil away from its parent thunderstorm is a function of the wind at the anvil level. In a jet stream situation, the anvil may get blown off more than 200 km away from the parent cloud. There have been instances over Kolkata when the anvil of a Cb cloud associated with a nor'wester has arrived some hours before the arrival of the actual nor'wester itself (see Section 2.3.3).

Although a thunderstorm may consist of just one ordinary cell that goes through the various stages of its life cycle and dissipates, thunderstorms very often form in clusters with numerous cells in various stages of development merging together. Unlike ordinary single cells, cluster storms can last for several hours producing large hail, damaging winds, flash flooding and isolated tornadoes.

If the air is conditionally unstable and there is a strong trigger, a line of several Cb cells, 150-250 km long, develops and is clearly seen in satellite images. This is called a squall line, because there are squalls ahead of it. Along this line, new storms may be forming and older storms may be existing at different stages of their life cycle. The rain-cooled air or gust front spreads out from underneath the squall line and acts as a mini cold front, continually lifting warm moist air to maintain the storm activity. Gusty winds flow out and spread horizontally along the ground. Squall lines are common over north and northeast India during winter and pre-monsoon months. Super-cell thunderstorms are a special kind of single cell thunderstorm that can persist for many hours. They are responsible for nearly all of the significant tornadoes produced in the U.S. and for the formation of giant hailstones.

Manohar et al (2003, 2004 and 2005) have produced a comprehensive climatology of thunderstorms over the Indian region including their spatial distribution, the number of thunderstorm days, the number of rainy days, etc. This has been further updated by Tyagi (2007). The pre-monsoon months of March-May, conditions are favourable for occurrence of thunderstorms as there is availability of moisture in the lower levels, and cold sub-tropical westerlies prevail in the upper levels. There are two prominent zones of thunderstorm activity, having 10-12 storms a month, one over the hills of northeast India and another over Kerala. Pre-monsoon thunderstorms are also common over other parts of northwest India, over northeast India, and to a lesser extent over the central and southern peninsula. During the monsoon season, the zone of thunderstorm activity shifts to the north and lies in the vicinity of the monsoon trough. During the post-monsoon and winter seasons, the number of thunderstorms over India is very small except over Kerala and the northeast.

Thunderstorms appear bright white in all channels of satellite imagery. In IR images, thunderstorm anvils can be seen as bright white oval-shaped cloud masses blown away in the direction of the upper winds. In a VIS image, the tall clouds may sometimes cast a shadow on the lower level clouds.

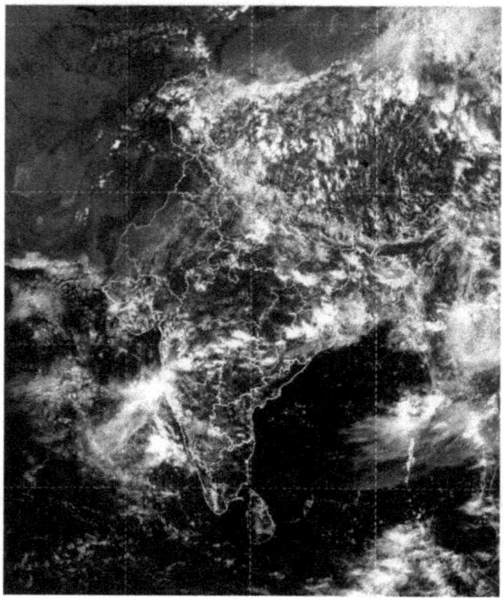

Figure 2.2.1.1 INSAT-3DS VIS image of 6 August 2025 0800 UTC showing the development of afternoon thunderstorms.
(Source: https://mausam.imd.gov.in/)

Figure 2.2.1.2 INSAT-3DS TIR1 image of 6 August 2025 0800 UTC showing the development of afternoon thunderstorms.
(Source: https://mausam.imd.gov.in/)

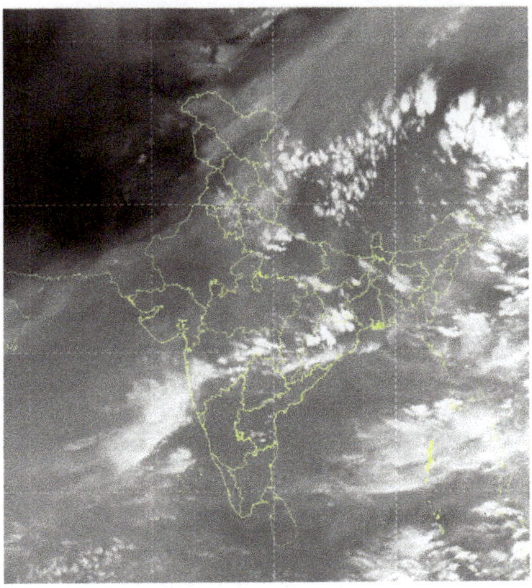

Figure 2.2.1.3 INSAT-3DS WV image of 6 August 2025 0800 UTC showing the development of afternoon thunderstorms.
(Source: https://mausam.imd.gov.in/)

Figures 2.2.1.1 to 2.2.1.3 show the development of afternoon thunderstorms over central India during a weak phase of the southwest monsoon. The mature thunderstorms appear bright white in the VIS, IR and

WV images because of the strong reflection from the Cb tops, low cloud top temperature and high moisture content respectively. Figure 2.2.1.4 shows spots of lightning flashes providing corroborative evidence of thunderstorm activity seen in satellite imagery.

Hail is a solid form of precipitation that is formed when updrafts in thunderstorms carry raindrops high up to extremely cold levels. Hailstones grow by collision with super-cooled water drops. In wet growth, the hailstone nucleus or an ice particle is in a region where the air temperature is just below freezing point. With dry growth, the air temperature is well below freezing and the water droplet immediately freezes as it collides with the nucleus. The stronger the updrafts within the thunderstorm, the larger the size to which a hailstone can grow. However, there eventually comes a stage when the updrafts can no longer match the weight of the hailstones and the precipitation falls to the ground in the form of hail.

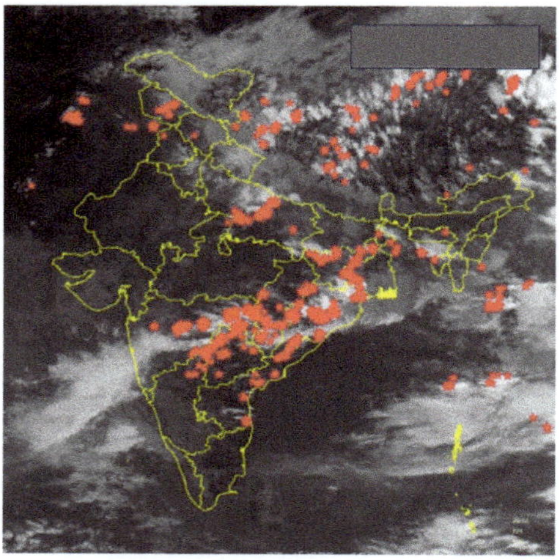

Figure 2.2.1.4 INSAT-3DS TIR1 image of 6 August 2025 0800 UTC superimposed with red dots showing the occurrence of lightning. (Source: https://mausam.imd.gov.in/)

There are only a few countries in the world that experience hailstorms frequently and India happens to be one of them. Hailstones comparable in size to lemons, mangoes or cricket balls have been reported in many instances of hailstorms in India. Hailstorms are predominant in the pre-monsoon and winter months over north and northeast India and Bangladesh, but they do not occur during the monsoon. Hailstones can destroy standing crops and may cause damage to vulnerable structures. Some of the hailstorms may be so intense that people and animals could even get killed if exposed directly to falling hail. In February and March 2014, an unusual interaction between cold northerlies and moist easterlies from the Bay of Bengal produced a series of severe hailstorms over central and western India. Crops in the state of Maharashtra in particular suffered unprecedented damage.

A cloudburst is a devastating weather phenomenon in which very heavy rainfall gets concentrated over a small area for a short time, leading to flash floods, structural collapses and human casualties. Although the mechanism of cloudbursts is not yet fully understood, it is generally believed that they are the manifestations of intense vortices which generate strong convective currents. These currents rapidly lift up the moisture laden air to form Cb clouds, particularly in mountain regions where cloudbursts occur more commonly.

Himachal Pradesh in India is one such region prone to cloudburst particularly in the monsoon season (Bhan et al 2004). A majority of them occurred in July-August, and Kullu and Shimla districts are the most vulnerable. A possible explanation is that some monsoon lows or cyclonic circulations after moving in a northwestward direction, recurve over Rajasthan or northwest Madhya Pradesh, and move on to Himachal Pradesh resulting in cloudbursts and landslides. The orography of the region is such that rain water from the cloudbursts rushes downslope, carrying debris, boulders and uprooted trees at great speeds and causing loss of life and property. In 2025, cloudbursts and landslides caused extensive damage to property and loss of human life in Uttarakhand, Himachal Pradesh and Jammu and Kashmir.

2.2.2 Dust-storms or *Andhi*s

When it has not rained over a place for a prolonged period, the soil becomes dry and the particles of the top soil layer get detached. In this situation, the occurrence of a thunderstorm results in large amounts of dust and sand getting thrown up into the air before the commencement of rain. If the area is excessively hot and dry, rain drops may get evaporated on their way down and may not even reach the surface, the result being a dust-storm.

Dust-storms are of two types, pressure gradient type and convective type. Over northwest India, particularly Rajasthan, convective dust-storms which are locally called *andhi*s, commonly occur in the summer months of March to June prior to the arrival of the southwest monsoon rains (Joseph et al 1980, Joseph 1982). These storms are characterized by downdrafts of thunderstorms or squall lines that are so intense that they throw up dust walls on the ground as far as 30 km distance from the main thunderstorm. Even after a dust-storm has ended, dust particles may remain suspended in the atmosphere for quite a long time, reducing visibility and producing haze. The horizontal visibility may get significantly reduced, even to 100 m. Dust-storms also result in a cooling of the atmosphere and give some respite from the summer heat. In India, dust-storms are classified as light, moderate or severe according to the wind speed and visibility level. Dust-storms usually occur in the afternoons, but they are not uncommon in the evening or night in which case they last for a longer duration. After the southwest monsoon has advanced into Rajasthan and the monsoon showers have moistened the soil, the frequency of dust-storms comes down drastically.

Dust-storms cannot be seen in IR images because of the lack of temperature difference between the surface and the dust layer. However, with a suitable enhancement, dust-storms can be detected in VIS images from geostationary satellites if they are of sufficient horizontal extent. MODIS images give an extremely clear view of the dust raised by dust-storms but only if there is a satellite pass at the time the storm has occurred.

Figure 2.2.2.1 Suomi NPP satellite VIIRS image of 22 January 2022 of a dust- storm on the Arabian Sea. (Source: https://www.ssec.wisc.edu/)

Figure 2.2.2.1 is an image acquired by the VIIRS instrument on the Suomi NPP satellite on 22 January 2022, when plumes of desert dust affected populated cities like Karachi and Mumbai, with degraded air quality and reduced visibility. The image shows dust plumes over western India and all over the Arabian Sea. This was an unusual situation as it occurred during the winter season because of an anomalous wind pattern.

2.2.3 Nor'westers or *Kal-Baisakhis*

In the pre-monsoon months, particularly April and May, violent thunderstorms develop regularly over the eastern and northeastern regions of India and

Bangladesh. Strong afternoon heating of the landmass initiates convection which intensifies as it extends towards the southeast and encounters warmer moist air. These thunderstorms are called nor'westers because they appear to arrive at a place from the northwesterly direction or locally as *kal-baisakhi*s, meaning calamity during the month of Baisakh of the Indian calendar.

Nor'westers have their origin over Jharkhand, Chhattisgarh or west Orissa and move eastwards or southeastwards towards Gangetic West Bengal at a speed of 50-60 km/hr. They do not advance much into the sea. Nor'westers are generally accompanied by squalls with wind speeds reaching even up to 150 km/hr.

Although the nor'wester is a mesoscale phenomenon, the realization of instability depends upon the large scale flow and the prevailing synoptic situation. During April-May, there is a shallow layer of moist southerlies and southeasterlies extending from the Bay of Bengal into Gangetic West Bengal. In the upper troposphere, however, there are cold and dry westerly winds associated with the sub-tropical jet stream or trough. This combination of moist southerlies below and cold westerlies aloft, results in the generation of the violent nor'westers.

Karmakar et al (2005) have studied the different energy components of the troposphere prior to the occurrence of nor'westers over Bangladesh. Their study covered more than a hundred nor'westers that occurred over Bangladesh during the period 1990-1995. They found that on the days the nor'westers occurred, both the sensible heat and latent heat increased significantly from the surface to 200 hPa level.

With the aim of an improved understanding of the thermodynamic features of the atmosphere as well as the forecasting skill of the occurrence of nor'westers, the Department of Science and Technology and Ministry of Earth Sciences of the Government of India launched a multi-institutional programme named Severe Thunderstorms Observations and Regional Modelling (STORM) over Gangetic West Bengal and northeast India in 2005.

In a study done under the STORM programme, Tyagi B. et al (2012) have delineated the characteristic differences in the thermodynamical structure of the atmosphere, convective and stability indices during days of occurrence and non-occurrence of thunderstorm activity during the pre-monsoon season of 2010 at Kharagpur using Doppler radar data. In another STORM investigation, Tyagi A. et al (2012) used INSAT VIS and IR imageries to document the convective cells which developed over east and northeast India and neighbouring Bangladesh, Bhutan and Nepal in the year 2009. Both the studies have demonstrated that INSAT satellite imagery together with overlapping radar data would be able to provide important clues for the detection of convective formations and their growth into mesoscale convective complexes over the nor'wester region.

2.2.4 Tornadoes

A tornado is a highly concentrated vortex of very small horizontal dimensions. It has a violent speed of rotation with wind speeds of the order of 300-400 km/hr around a vertical or near-vertical axis, a strong updraft and a very high pressure deficit within it, even 50-100 hPa. It has an extremely short life cycle of barely 10-20 minutes. The tornado is visible in the sky as a narrow funnel tapering from the base of a thundercloud. The end of the funnel may touch the ground or it may remain suspended in the sky. When the funnel is over a water body, it raises water vertically in the form of a pillar, and it is then known as a water spout. Over land, it can cause extreme devastation. About one thousand tornadoes strike the U.S. each year. Oklahoma City has been one particular place that has historically been the worst hit. In the biggest outbreak of tornadoes,147 tornadoes touched down in 13 U.S. states on 3 and 4 April 1974. Hurricanes are also known to generate tornadoes within their outer bands.

Tornadoes are relatively rare over the Indian sub-continent. During the 160-year period from 1839 to 1999, only 64 tornadoes are known to have occurred in this region (Golder et al 2001). On 17 March 1978, in the northern suburbs of Delhi, near the Delhi University, 28 people were killed and 700 were injured by a tornado that cut a path 5 km long and 50 m wide (Gupta et al 1980). On 24 March 1998, a violent tornado killed 160 people and injured 2000 when it tracked through 20 coastal villages in the eastern states of West Bengal and Orissa. As tornadoes are extremely localized events, it is quite possible that some cases might not even have got reported. The area of northeast India and Bangladesh is prone to the development of tornadoes. They generally occur in the pre-monsoon months of March-May, and more commonly in April. The favourable time of occurrence is afternoon or evening.

A severe thunderstorm produced a tornado (F3 on the Fujita-Pearson scale), which affected Rajkanika block of Kendrapara district of Orissa in the afternoon of 31 March 2009. It took 15 lives and left several injured and destroyed a lot of property. The meteorological conditions that led to this tornado have been analyzed by Litta et al (2012). There was a similar instance close to Ludhiana airport in Punjab, on 15 August 2007. The damage was relatively less as the tornado passed through the open fields (Litta et al 2020).

Tornado development can be spotted in satellite images by continuous monitoring especially when the synoptic situation is known to be favourable, but otherwise they can be missed because of their small size and short life. Satellites cannot directly detect tornadoes as they form close to the ground, often beneath thick cloud cover. They can however detect overshooting cloud tops and above-anvil cirrus plumes that signal strong updrafts and act as precursors to tornadoes.

On 24 May 2017, the GPM core observatory satellite flew above a line of storms spawning tornadoes that were moving across Florida. The GPM radars collected data showing the very heavy downpours accompanying these storms, at a rate of over 215 mm per hour (Figure 2.2.4.1).

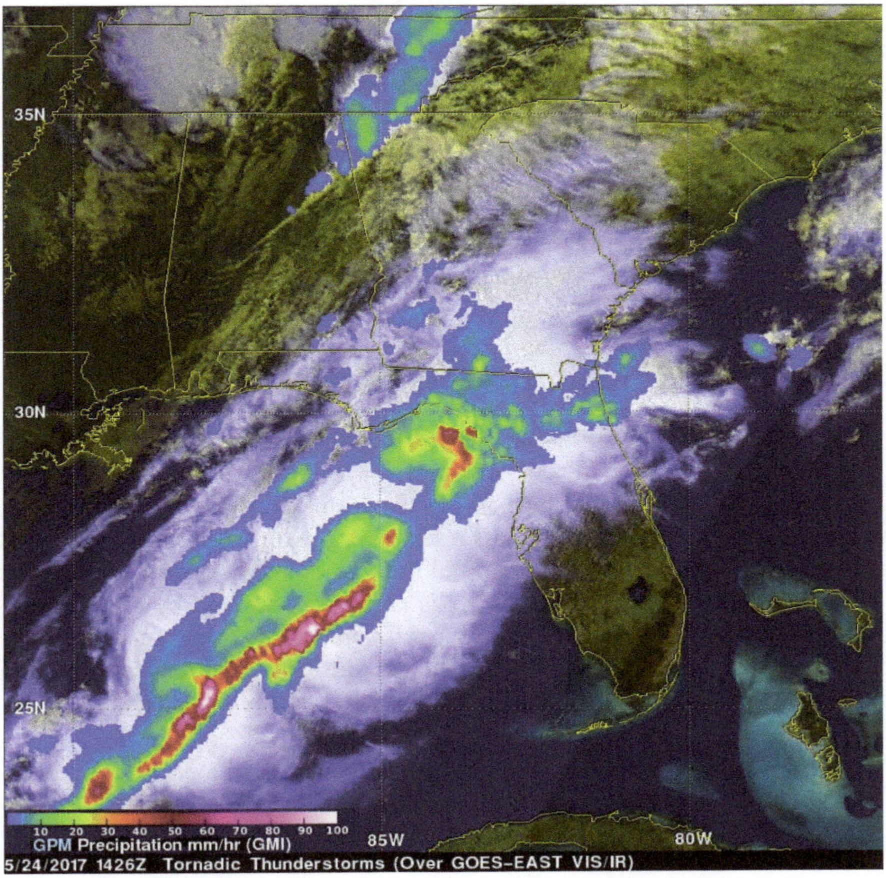

Figure 2.2.4.1 GOES VIS-IR image superimposed with GPM's Microwave Imager showing tornado-associated rainfall on 24 May 2017 over Florida. (Source: https://gpm.nasa.gov/extreme-weather/)

2.2.5 Mountain Waves

When the wind is blowing perpendicular to a mountain range, and the atmosphere has thermally stable layers, waves form on the leeward side of the mountains producing strong turbulence and vertical air currents. After the winds cross over the mountain tops, air in the lower layers descends into the valley, and rises again to form what is called a rotor cloud. At the higher levels, an internal lee wave builds up above the rotor cloud (Figure 2.2.5.1). The character of the wave varies with the prevailing wind profile. When the jet stream is favourably aligned to the mountains, the stronger winds may give rise to more intense waves.

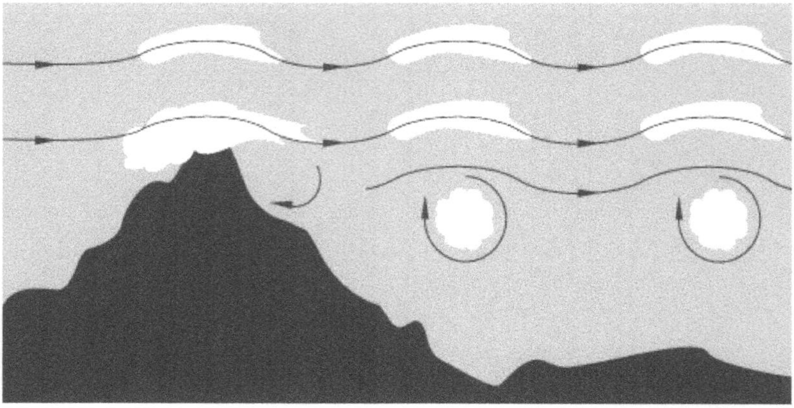

Figure 2.2.5.1 Schematic of formation of mountain waves.

If there is sufficient moisture supply, orographic clouds also develop and are identifiable in satellite images as a series of gray or white cloud bands with intervening clear areas running parallel to the mountains, sometimes even up to several hundred km. Figure 2.2.5.2 is an image acquired by the Landsat-8 Operational Land Imager (OLI) on 2 November 2014. It shows mountain wave clouds flowing in a southeasterly direction on the lee side of Amsterdam Island in the southern Indian Ocean, seen on the far left part of the image.

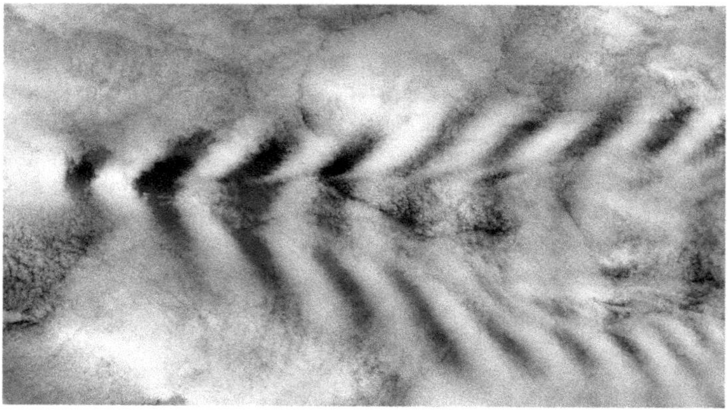

Figure 2.2.5.2 Satellite image of cloud bands in association with mountain lee waves. (Source: NASA web site earthobservatory.nasa.gov)

Mountain waves have been a subject of considerable study particularly because they are a potential aviation hazard. They create strong turbulence and downdrafts on the leeward side of the mountains which can be dangerous to flying. There have also been several studies in which lee waves identified in satellite images have been compared with theoretical computations of their wavelengths. De (1970, 1971) reported satellite evidence of lee waves over northeast India and compared the observed wavelengths with theoretical estimates using a two-dimensional steady state linear model.

In the winter season, the air stream over northwest India is stably stratified and is favourable for the formation of lee waves provided the air flow has an appreciable component normal to the mountains. Sinha Ray (1988) who used NOAA satellite imagery for the years 1982-85 for the western Himalayas, found that the wavelengths were between 10 and 30 km and their computed values were in good agreement with those observed in satellite images.

2.3 Inter-Tropical Convergence Zone

A very conspicuous and prominent feature of geostationary satellite images or global mosaics constructed from successive passes of polar orbiting meteorological satellites is the Inter-Tropical Convergence Zone (ITCZ). Satellite imagery shows up the ITCZ as a complex cloud band encircling the earth. Of course, long before the satellite era began, the ITCZ was well-known as a zone of convergence of air masses between the two hemispheres. In the lower troposphere there is a wind discontinuity with horizontal velocity convergence and net upward motion. This results in an east-west band of cloud and rain which oscillates north-south as per the position of the sun.

Asnani (2005a) gave a detailed description of the characteristic features of the ITCZ and the large geographical and seasonal variations in its location. Over regions of eastern Pacific and Atlantic Ocean, it remains north of the equator throughout the year. In other regions, it oscillates between latitudes north of the equator in the northern summer and south of the equator in the southern summer.

From west Africa to southeast Asia, the discontinuity is between the westerlies in the near-equatorial region and the easterly trade winds on either side of it. The westerlies are largely the southeast trade winds which have changed direction after crossing the equator. Over the Atlantic and Pacific Oceans, the discontinuity is between the northeast and southeast trades of the two hemispheres.

Since the ITCZ has a tendency to be located over the warmest regions of land or sea, it is also sometimes called the thermal equator. The ITCZ is an important mechanism that transports heat from the tropical planetary boundary layer to the middle and upper troposphere. The upper limb of the Hadley cell carries it to the sub-tropical latitudes from where it gets further transported to higher latitudes.

Although the ITCZ is broadly regarded as the region of convergence of the trade winds from the northern and southern hemispheres, it must be understood that the air masses on either side of it are not always from the respective hemispheres. As a case in point, during the Indian southwest monsoon season, the origin of the air mass just south of the ITCZ cannot be always be traced back to the southern hemisphere.

The cloud band characteristic of the ITCZ does not appear permanently around the equator, nor does is it take the form of a continuous belt. Due to associated easterly waves and mesoscale circulations, the ITCZ appears in satellite pictures as a series of cloud clusters interspersed with clear areas, as for example, in the Kalpana-1 full disc image of 1 May 2005 (Figure 2.3.1).

Figure 2.3.1 Kalpana-1 IR image of 1 May 2005 showing the ITCZ clouds extending across the Indian Ocean. (Source: IMD)

In one of the earliest satellite-based studies of equatorial cloudiness over the Indian Ocean, Saha (1971) observed many typical features. The equatorial Indian Ocean within 5° of the equator, has much more clouding east of the 60 °E longitude than to its west throughout the year. Clouds in the equatorial Indian Ocean tend to form in clusters rather than bands. The cloud clusters are randomly distributed with respect to the equator, but occasionally there are two major cloud bands one on either side of the equator. During the northern summer, the cloud band may be located as far north as 20-25 °N, while in other times of the year, the bands are within 10-15° of the equator. These observed features can be explained in terms of the boundary level convergence due to the thermal wind effect.

A satellite-derived climatology of the ITCZ based upon satellite VIS and IR channel data from 1971 to 1987 was compiled by Waliser et al (1993). It gives global monthly mean positions of the ITCZ as well as its structural characteristics. The ITCZ clouding is characterized by a narrower well-defined band over the Atlantic and eastern Pacific Oceans, while it occupies a broader latitude belt over the Indian and western Pacific Oceans. It is broad and also irregular over land regions of the tropics. An interesting feature of the satellite climatology is that cloudfree areas over the equator are not always associated with colder waters, and even with high sea surface temperature, deep convection does not necessarily develop.

The clouding associated with the ITCZ occasionally takes the form of a double cloud band on either side of the equator, as seen for example, over the Indian Ocean on 24 June 2006 (Figure 2.3.2). This type of double cloud band and the causes of subsidence near the equator have been the subject of an extensive analysis and discussion by Asnani (2005b). He regards the equator not just as a geometric line, but as having dynamical effects on the flow of a fluid crossing it. He has postulated that air approaching the equator from north or south suffers horizontal velocity divergence, and air going away from the equator towards north or south suffers horizontal velocity convergence. When this occurs in the lower troposphere, the air approaching the equator descends, while the air leaving the equator ascends. The equator thus acts like a dynamical valley. This is borne out by the observed tendency of highly reflecting clouds to avoid the geographical equator. According to Asnani, the ancient perception that the equator is a region of abundant cloudiness and perpetual rain is now obsolete, and it would be more correct to look upon the equator as a region of minimum cloudiness instead.

Figure 2.3.2 Kalpana-1 IR image of 24 June 2006 showing double cloud band of the ITCZ over the Indian Ocean. (Source: IMD)

2.4 Monsoons

Monsoons have been defined in various ways in terms of alternating distinct wet and dry seasons during the year, a seasonal reversal of wind direction and the north-south oscillation of the ITCZ, which itself is a region of wind discontinuity. The monsoon is essentially an annual oscillation of the state of the atmosphere in response to the changing position of the sun between the tropic of Cancer and the tropic of Capricorn. The traditional view of the Asian monsoon was that of a giant land-sea breeze phenomenon caused by the differential heating between the Asian land mass and the Indian Ocean.

The monsoon region was originally considered to be restricted to south and south-east Asia, northern Australia and tropical Africa. Around the 1980's, the summer rains in Mexico and south-west U.S. also came to be classified as monsoon rains. The current thinking is that the monsoons are an important part of the global large-scale circulation and the whole tropical region and adjoining sub-tropical region is considered as the monsoon region, which includes parts of north, central and south America, China and east Asia (Kelkar 2012).

This does not mean, however, that the ITCZ moves to 40° N, but there are changes in the intensity and position of the sub-tropical trough lines which are associated with the monsoon in China and east Asia, south Pacific Ocean and south Atlantic Ocean. Then there are some regions like sub-tropical south America which exhibit monsoonal characteristics during winter rather than in summer.

In essence, the monsoon is where the ITCZ is (Asnani 2005c). The monsoon is a fast shift of the ITCZ away from the equator. As far as the south Asian monsoon is concerned, the monsoon oscillation is stronger in the northern hemisphere than in southern hemisphere and it is stronger over south and south-east Asia than elsewhere in the northern hemisphere. This can be attributed to the Himalayan mountains and elevated Tibetan plateau producing diabatic heating over a large area of the middle troposphere, the Indian Ocean to the south providing abundant moisture supply and the strong meridional gradients of temperature.

Rajeev et al (2006) have constructed monthly mean fields of fractional cloudiness observed by INSAT and NOAA satellites over the years 1996 to 2003. The annual north-south march of the ITCZ, the intense cloudiness over south Asia during the southwest monsoon months of June-September, and its near absence in other seasons, are clearly brought out in the monthly cloudiness fields. The intra-seasonal behaviour of the monsoon is, however, equally important, and satellite observations have been of great help in understanding it as will be explained in the following sections.

2.4.1 Onset and Advance of Southwest Monsoon

The onset of the southwest monsoon over Kerala is a part of a large-scale circulation event and it is associated with slowly evolving as well as sudden changes in many atmospheric and oceanic parameters like low level winds, moisture, upper level temperature and SST (Joshi et al 1994, Joseph et al 1994, Khole et al 2004). In early April, the tropical north Indian Ocean and adjoining west Pacific Ocean begins to warm rapidly and by the end of May, it becomes the warmest tropical ocean area. After the onset of the monsoon, it gets cooled by the strong monsoon westerlies. Convective activity over north Indian Ocean and cyclogenesis over west Pacific Ocean are components of the 30-50 day cycle observed over the areas and there is a see-saw variation between the two, which influences the onset of monsoon over Kerala (Joseph 1990).

The onset of monsoon over Kerala has been the subject of numerous investigations (Joseph et al 2003, Pearce et al 1984, Shenoi et al 1999, Soman et al 1993) and of particular interest has been the generation of the so-called onset vortex that develops over southeast Arabian Sea (Mishra et al 1991, Krishnamurti et al 1984, Rao 1990, Rao et al 1999). However, it is the arrival of the monsoon rains over Kerala which is the prime consideration in defining the onset, because it paves the way for their further advance into the interior parts of India. The monsoon onset in south Bay of Bengal and Myanmar takes place in May. In June, monsoon covers the Bay of Bengal, and extends into southwest China, Arabian Sea and the South China Sea (Wang et al 2004). By the end of June, the whole rain belt in the area stretching from the Arabian Sea to the South China Sea and further eastward, is well entrenched. The mean onset date over Kerala is 1 June with a standard deviation of a week. The earliest ever onset has occurred on 11 May (1918, 1955) and has been as late as on 18 June (1972).

The subsequent advance of the monsoon into the interior parts of the country is a slow process and the monsoon normally reaches extreme west Rajasthan around 15 July. There are major exceptions, however, like 2015 in which the monsoon covered the entire country as early as on 26 June. The standard deviation of the onset dates over other places in India is about 6 to 8 days.

There have been many attempts to arrive at an objective definition of the onset of monsoon over Kerala (Ananthakrishnan et al 1988, Fassulo et al 2003, Joseph et al 2006) but there is a divergence among meteorologists on this issue. In the pre-satellite years, the onset over Kerala was defined in terms of a detectable change in the pentad rainfall pattern over Kerala. This approach does not always determine a clear onset since there is considerable thunderstorm rainfall before the monsoon onset. There are other precursors of the onset such as a ssteady increase of convection over an area bounded by 0-15°N and east of 60°E about 3 pentads before onset and a steady acceleration of the cross-equatorial Findlater Low Level Jet (LLJ) in response to this heat source. There is an increase in the depth (600 hPa) and strength of the westerlies in the LLJ just prior to onset. The axis of the LLJ also moves from the equatorial Arabian Sea to higher latitudes in the 3 pentads before onset.

The characteristic features associated with the onset over Kerala are generally known, but all of them may not always be present. It is even more difficult to define the monsoon onset or advance into interior areas of India. For the common people, particularly farmers, onset must signify sustained rains. Very often the onset of monsoon over Kerala is accompanied with thunder, heavy rain, strong winds, rough seas, etc. Westerlies are strong and surface wind speeds are of the order of 20 knots. The westerlies extend aloft up to 6 km and rainfall is of the order of 10 cm in a day. In some years, the onset is of a subdued nature.

The north-south sea level pressure gradient over India exhibits an annual see-saw oscillation. Between November and April, the pressure decreases from north to south. Between May and October, the pressure increases from north to south. In the transition months, the gradient is almost zero.

In the pre-monsoon season (March-May), there is considerable thunderstorm activity over the country. Kerala and northeast India have on an average 10-12 thunderstorm days per month during March-May. The first storms may be dust-storms due to the rising of loose soil particles after a long period of dry weather. Some thunderstorms may have squally winds and hail. After the monsoon has established itself, the number of thunderstorms reduces considerably.

The best precursors of monsoon onset over Kerala are observed in satellite imagery such as that provided by INSAT. In many years, INSAT images show convective clouds along the west coast of India, indicative of a trough in the westerly winds over the Arabian Sea. Figure 2.4.1.1 shows a Kalpana-1 image of a weak and delayed onset on 5 June 2005 with no clouds over the country except over the southern tip. Figure 2.4.5.1 shows the satellite image for 26 May 2006 when there was an early and strong monsoon onset over Kerala with an associated onset vortex. WV imagery vividly brings out the onset over Kerala and shows the distinctly moist air of the monsoon with dry regions to its south and north (Figure 2.4.1.2).

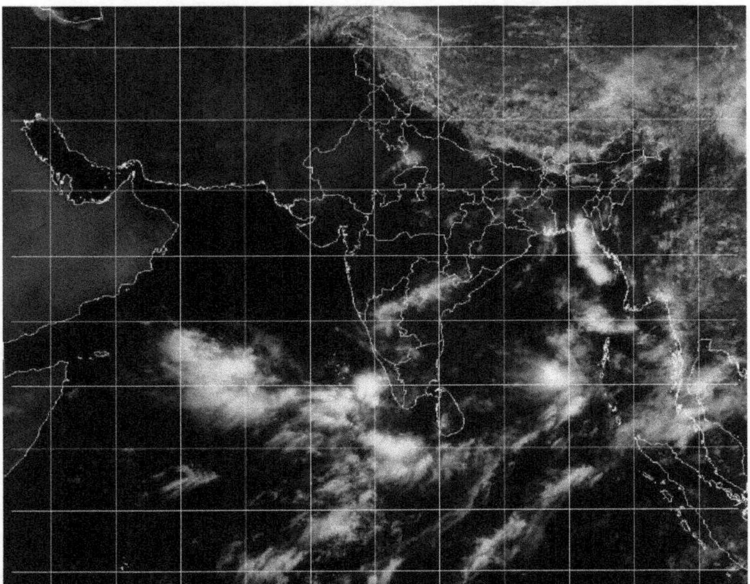

Figure 2.4.1.1 Kalpana-1 VIS image of 5 June 2005 showing a weak and delayed onset of the southwest monsoon over Kerala. (Source: IMD)

Weekly data of Oceansat-1 MSMR for the pre-monsoon season of the year 2000 showed that as the monsoon progressed towards the Indian sub-continent, the area of maximum moisture content and wind speed moved northwards. There was a sharp increase in water vapour over western Indian Ocean about three weeks prior to the onset over Kerala and a rise in the wind speed over western Arabian Sea just prior to the onset (Simon et al 2001).

Figure 2.4.1.2 Kalpana-1 full disc WV image of 27 May 2006 showing the onset of the southwest monsoon, and the distinctly moist air of the monsoon with dry regions to its south and north. (Source: IMD)

Pai and Rajeevan (2009) have described the new criteria adopted by IMD in 2006 for declaring the onset over Kerala operationally. They are based on rainfall, wind field and OLR. If after 10 May, 8 out of 14 selected stations report rainfall of 2.5 mm or more for two consecutive days, onset may be declared on the second day, provided certain specific wind and OLR criteria are also satisfied in concurrence.

2.4.2 Low Level Jet

By definition, a jet stream in the atmosphere is a strong narrow current, running across thousands of kilometres, with a width of a few hundred kilometres and a depth of at least a kilometre (Berggren et al 1958). Jet streams have wind speeds greater than 60 knots concentrated along a quasi-horizontal axis and the show strong wind shears of the order of 10-20 knots/km in the vertical direction and 10 knots/km in the horizontal direction. They are generally characterized by one or two vorticity maxima. There are, however, some currents in the atmosphere which do not strictly meet these criteria but still qualify to be called jet streams. One such current is the Low Level Jet (LLJ) which develops along the east African coast in association with the Indian southwest monsoon. This phenomenon, which was first reported by Bunker (1965) and fully explained afterwards by Findlater (1966, 1969), has since come to be called the Findlater Jet. Some characteristic features of the LLJ have been discussed by Pant (1982) and Joseph et al (2004).

The LLJ which accumulates moisture from the sea surface is now regarded as a significant factor in the cross-equatorial moisture transport that is important to monsoon dynamics. It has its origin in the southern hemisphere and in the month of July, the current flows from Mauritius over the northern tip of Malagasy into the coast of Kenya in east Africa at around 3 °S. Thereafter it penetrates inland into Kenya, Ethiopia and Somalia and emerges into the southwest Arabian Sea at 9 °N and moves towards the west coast of India. The LLJ remains a narrow stream along the African coast because of the mountains there, but orography also helps the wind to intensify near the surface and run parallel to the coasts of Somalia, Yemen, and Oman.

The LLJ is remarkable for its steadiness of direction but not necessarily in its strength. At the 850-900 hPa level, speeds of 30 knots are generally observed over Malagasy and 35 knots near the Somali coast in July and August. However, much higher speeds are occasionally encountered, Findlater (1974) having reported an extreme value of 126 knots.

The Somali current off the Somali coast, is an inertial boundary current forced by the atmospheric LLJ. As the LLJ interacts with the waters off the Somalia coast, it produces intense upwelling and cold water rich in nutrients is brought up to the surface. This leads to convergence and downwelling in the central and eastern part of the Arabian Sea. In response to the southwest monsoon, the surface mixed layer in the central Arabian Sea deepens and cools during the summer. The cross-equatorial LLJ is a major component of the lower limb of the monsoon circulation connecting the southern Indian Ocean and the south-east Asian region. Intensification of the LLJ usually leads to an increase in the monsoon activity over west coast of India (Ngara et al 1978).

Generally, low level jets are strongly influenced by factors such as orography, friction, diurnal heating cycle and variations of pressure gradient. As they have limited horizontal and vertical extent, they may not always show up in mean climatological charts. Using aircraft flight data gathered during MONEX-79, Grossman et al (1986) have demonstrated that the structure of the east African LLJ can be explained by a combination of the geostrophic wind shear or the thermal wind, above the wind maximum and turbulent momentum transfer below it.

Joseph (2006) has attributed the LLJ with two main functions. One, it acts as a conduit carrying the moisture generated by the trade winds over the south Indian Ocean and the evaporative flux from the Arabian Sea to the areas of monsoon rainfall production over south Asia. Two, the cyclonic vorticity north of the LLJ axis in the atmospheric boundary layer serves as a dynamic forcing for the generation of upward air motion, convective clouds and associated rainfall and for the generation of monsoon depressions in the north Bay of Bengal. During an active monsoon, the core of the LLJ passes eastward across peninsular India, while during monsoon breaks, the LLJ axis moves southwards from the central Arabian Sea, skirts India and passes eastward.

Satellite-derived cloud motion winds over the Indian Ocean, particularly those derived with low level clouds, clearly bring out the LLJ (Figure 3.1.3.1).

2.4.3 Tropical Easterly Jet

During the summer monsoon season, marked north-south temperature and pressure gradients develop between the Indian Ocean and the Tibetan plateau. In the lower troposphere, pressure increases equatorward and in the middle and upper troposphere, pressure increases poleward. As the Coriolis parameter is small in the lower latitudes, even a slight pressure gradient can give rise to strong easterly flow with speeds going up to 60 knots at 100-300 hPa levels. This is manifested as the Tropical Easterly Jet (TEJ). TEJ extends from southeast Asia to north Africa in the latitude belt 10-15 °N. The strongest winds occur just to the west of the southern tip of India, over the Arabian Sea at about 150 hPa. The TEJ is very stable in direction but the wind speed is variable. The vertical wind shear is of the order of 10 knots/km and the horizontal wind shear is very weak, 10 knots/500 km in contrast with the sub-tropical westerly jet (10 knots/100 km).

The TEJ is an important feature of the southwest monsoon and is easily identifiable in satellite imagery (Figure 2.4.5.1). Singh et al (1982) made one of the earliest studies of TEJ using satellite pictures during June 1977. The TEJ is more evident in IR images than in VIS images because of the associated cold cirriform clouds. Extended cirrus bands parallel to the jet axis are observed, sometimes with a sharp equatorward edge indicative of descending air over the equatorial region (Koteswaram 1958). Transverse lines were also observed over the Arabian Sea. The monsoon was seen to be active in the right entrance sector of the jet clouds.

The strong winds of the TEJ are also seen in high level satellite-derived cloud motion winds (Figure 3.1.3.2).

2.4.4 Monsoon Depressions and MTCs

During the monsoon season, a series of depressions form over the north Bay of Bengal. They generally move in a west-northwesterly direction across the country and often merge into the seasonal low over northwest India. Some of them take a westerly course and in rare cases even emerge into the Arabian Sea. Some may take a north-northeasterly course and dissipate over the mountain slopes, causing flooding of rivers which originate there. If depressions linger over central India, it is the peninsular rivers which get flooded.

Generally, 4 to 6 monsoon depressions form every year but the number is highly variable. In the 2005 monsoon season, there were as many as 11 low pressure systems, 5 of which intensified into a depression and 1 into a marginal cyclonic storm. There have also been other years when only a single depression had formed.

The time interval between two successive depressions is also variable. Depressions may follow in quick succession or may be after a long interval, which may signal the end of a break in the monsoon.

Most of the monsoon depressions typically have a life of 3 to 5 days, but some may last even for a week. When the seasonal trough intensifies and a closed low appears, a monsoon depression is likely to grow out of it. The development may occur *in situ* or be triggered by the arrival of an easterly wave. In a day or two, the low intensifies into a depression with winds of 22-33 knots. Depressions move at an average of about 4° latitude/longitude in a day but if they undergo a recurvature, then they move very fast. In many years, the advance of the monsoon over northeast India takes place is in association with a depression.

Unlike tropical cyclones, there is considerable asymmetry in the wind field of a monsoon depression. Winds are stronger to the south of the depression centre, due to the superimposition of the depression perturbation over the westerlies south of it. Winds are strongest at 700 hPa level. The system has a cold core below 700 hPa and a warm core aloft. The vertical axis of the depression slopes southwestwards with height. The heaviest rainfall occurs over the southwest quadrant because of convergence. Rainfall amounts are generally of the order of 10-20 cm per day and could even be as high as 30 cm per day in isolated cases. Rainfall in the southwest sector of the depression has considerable diurnal variation, with the heaviest falls occurring in the early morning hours.

There are three factors which are responsible for causing upward motion and rainfall in the southwest sector of a monsoon depression, (i) boundary layer frictional effects in areas of cyclonic vorticity which is maximum in the south sector, (ii) warm air advection in the lower troposphere and (iii) vorticity advection in the lower troposphere which contributes to upward motion in the left forward sector and descending motion in the rear sector.

In its early stages, a monsoon depression induces a low level trough along the Konkan coast, where there is enhanced rainfall activity, which decreases as the depression moves inland. Rainfall activity is high to the west of the depression, but less over southern India and northeast India. When the depression is over central India, Gujarat and Saurashtra get good rainfall because of fresh moisture supply from the Arabian Sea.

A monsoon depression in its formative stage is seen in satellite imagery as a dense amorphous cloud mass over the Bay of Bengal. As it develops further, convective cloud lines tend to get organized and there are pronounced curved cumulus lines at the edge of the dense overcast mass. In the mature stage, the dense overcast gets ragged edges surrounded by curved cumulus lines (Figures 2.4.4.1 and 2.4.4.2).

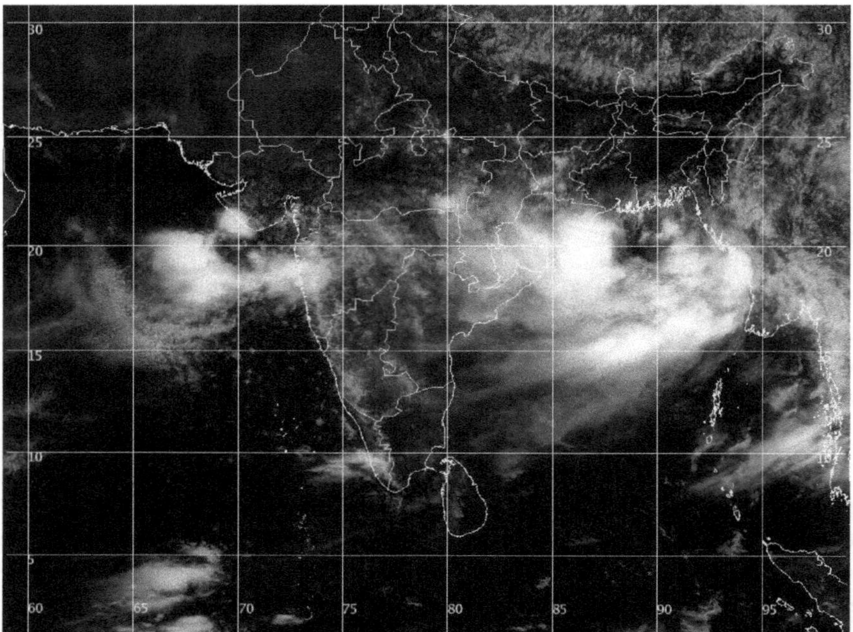

Figure 2.4.4.1 Kalpana-1 VIS image of 1 July 2006 showing the formation of a monsoon depression over the Bay of Bengal. (Source: IMD)

The fundamental characteristic features of monsoon depressions have been described by Srinivasan et al (1971), Chowdhury et al (1983) and Sarker et al (1988).

Chowdhury et al (1985) prepared a mean cloudiness pattern of a monsoon depression by examining 40 case histories during 1977-1981. A moving 20° × 20° lat/long grid was overlaid around the depression centre in the satellite picture and the cloudiness information picked up at each 2.5° grid point which was then averaged to draw isopleths of cloudiness and coefficient of variation. The mean pattern shows a zone of maximum cloudiness 3-4° southwest of the depression centre which could be identified as the principal area of heavy rainfall. The north and southeast extremity of the depression field is an area of suppressed cloudiness. The most distinct feature in all cases is the bright central dense overcast southwest of the depression centre and consisting mainly of stratiform clouds. In 50 % of the cases, the centre lay within this amorphous mass, but in the remaining cases the centre was located in the cloudfree zone indicative of a highly sheared environmental flow.

Prasad et al (1990) used 3-hourly INSAT images and other data to construct a composite wind and cloud structure of a monsoon depression. They have also described the cloud features seen in satellite images at different stages of development of the depression. The genesis stage which could be called a well-marked low, is characterized by a long east-west band south of the evolving vortex. In the intermediate stage at which the system becomes a depression, this zonal cloud band reorganizes itself suggesting a rotation of the cloud mass, while cumulus lines appear in the northern sector in the region of low level easterlies and define the centre. The centre could be 1-2° away from the centre fixed on the surface chart. In the intense or deep depression stage, the cumulus lines become more curved and the centre comes closest to the northern edge of the dense overcast. The vertical tilt of the system is the least at this stage. The dense overcast will have a 3-5° diameter with sharp edges and embedded overshooting cloud tops that penetrate the tropopause. The system begins to decay 24 hours later.

Mahajan (2003) and Mahajan et al (2004) have used a variety of satellite data from DMSP SSM/I and Oceansat-1 MSMR and cloud motion winds for studying the development of monsoon disturbances. They found that prior to the formation of a monsoon depression, there is a strengthening of surface winds to the south of the low pressure area over the Bay of Bengal. Highest values of geophysical parameters like integrated water vapour (6-8 gm cm^{-2}), integrated cloud liquid water (50-60 mg cm^{-2}) and precipitation rate (20-25 mm/hr) and lowest values of OLR (120 w m^{-2}) are associated with the southwest sector of a monsoon depression.

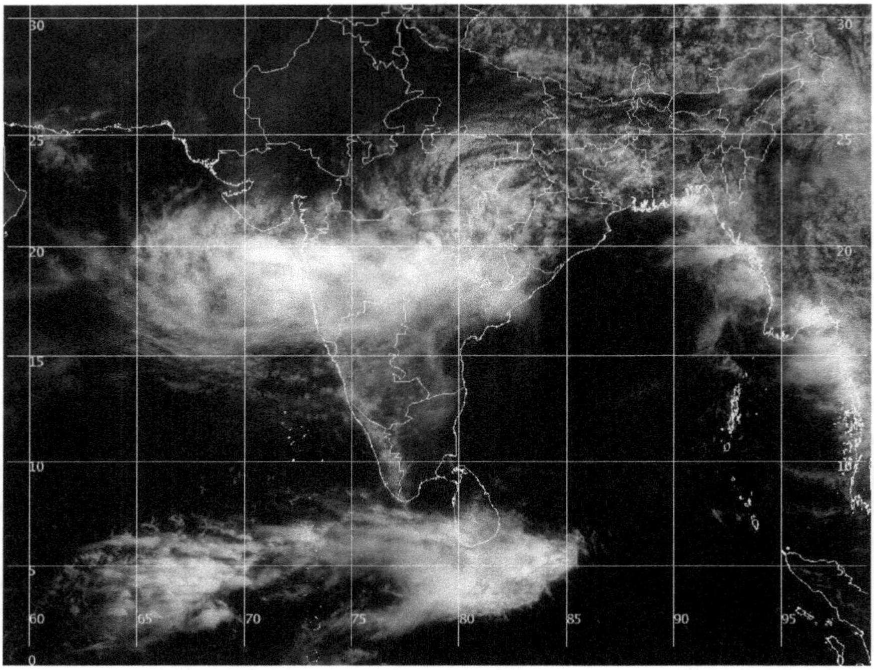

Figure 2.4.4.2 Kalpana-1 VIS image of 4 July 2006 showing a monsoon depression which has crossed from the Bay of Bengal into Orissa. The curved cumulus lines are seen. (Source: IMD)

Rajeevan et al (2001) examined 10 cases of monsoon depressions during 1989-1999 using SSM/I microwave data on daily water vapour, cloud liquid water and rain rate on a 25 × 25 km resolution. However, these data are available only on the ocean. They prepared a composite average pattern for a monsoon depression with respect to a common origin which confirmed the asymmetric distribution of precipitation. While atmospheric water vapour did not show any preferred maxima, there was high cloud liquid water (> 1 mm) in the southwest sector about 200 km from the common centre, which indicated the presence of very deep convective clouds. There the precipitation rate exceeded 12-14 mm/hr, equivalent to 30 cm rain in 24 hours. A secondary maximum was located in the northeast sector with rain rate of 6 mm/hr. Two case studies using TMI data revealed similar features.

A Mid-tropospheric Cyclone (MTC) is a synoptic system that is observed over South Asia during the summer monsoon months. An MTC is a low pressure system having a vorticity maximum in the middle tropospheric levels between 700 and 500 hPa, with much smaller values at the surface or at 200 hPa. It is, therefore, barely noticeable on surface weather charts, but produces heavy rainfall. MTCs are mostly warm core above 600 hPa, but a little cold core below (Miller et al 1968, Krishnamurti et al 1970). Bhagat (2005) computed some aspects of kinetic energy for the average MTC derived from 27 cases during 1985-1980.

Compared with monsoon depressions, MTCs are not that common. Some MTCs form *in situ* over the northeast Arabian Sea and adjoining land areas of Saurashtra. Others form out of monsoon depressions that have travelled from the Bay of Bengal across the country and emerged into the Arabian Sea. MTCs are relatively more frequent in the first half of the monsoon season. They have a tendency to remain quasi-stationary for several days or drift slowly westwards, unlike monsoon depressions which move as fast as 300-500 km per day.

2.4.5 Active and Weak Phases

During the southwest monsoon season, rain does not keep falling continuously all over the Indian subcontinent, but there are spells of vigorous monsoon rainfall interspersed by dry spells or periods of weak rainfall activity. This strong intra-seasonal oscillation that the monsoon exhibits is also referred to as the active-break cycle. In the active monsoon phase, which may last for 2 to 4 weeks, most parts of India receive good rainfall. During the break phase, however, rainfall is highly deficient except over the Himalayan foothills and extreme southern parts of India. Rajeevan et al (2006) have made an analysis of the active-break spells of the monsoon using high-resolution daily gridded rainfall data of the Indian region.

In one of the earliest studies of the monsoon based upon satellite imagery, Sikka et al (1980) studied daily variations of the maximum cloud zone (MCZ) and the 700 hPa trough over the region covering the Indian longitudes 70-90 °E from the equator to 35 °N in the months of April to October during the years 1973-1977. They observed that in the monsoon months of June-September, two MCZs exist in the region, one over the continent and another over the ocean. Both the MCZs exhibit alternating episodes of strong and weak intensity, but the strength of the two MCZs is oppositely correlated. Both the MCZs show a northward propagation, and there is a period of 3 to 4 weeks between successive generations of the northward moving cloud bands.

The monsoon trough is the only synoptic scale system that contributes to the rainfall activity over India. In satellite images the presence of cloud clusters helps to identify areas of heavy rainfall. Satellite images clearly bring out the weak phases of the monsoon followed by a revival and build-up. Figure 2.4.5.1 shows the satellite image for 26 May 2006 showing an early and strong monsoon onset over Kerala with an associated onset vortex. There was, however, a mid-way halt to its progression beyond the peninsula and a weakening of the monsoon current, but on 21 June there was a subsequent revival.

Figure 2.4.5.1 Kalpana-1 IR image of 26 May 2006 showing an early and strong onset of the southwest monsoon over Kerala. (Source: IMD)

Satellite pictures have confirmed that the cross-equatorial flow over the Indian Ocean has a predominant influence on the monsoon activity. Mishra et al (1991) and Gupta et al (1991) have highlighted the importance of the southern hemisphere equatorial trough (SHET) in regulating the cross-equatorial flow thereby leading to the development of the active-break cycle. They found an inverse relationship between the SHET clouding and the monsoon activity over India and persistence of the SHET cloud band beyond the pre-monsoon months could be a signal of an impending drought. Figure 2.4.5.2 shows the reappearance of a prominent SHET in the INSAT-3DS VIS full disc image of 29 July 2025 during a weak phase of the monsoon.

Figure 2.4.5.3 shows the southwest monsoon in an active phase once again on 3 July 2006. There is a monsoon depression over eastern India, flanked on either side by areas of intense convection on the Arabian Sea and Bay of Bengal, and an off-shore trough along the west coast.

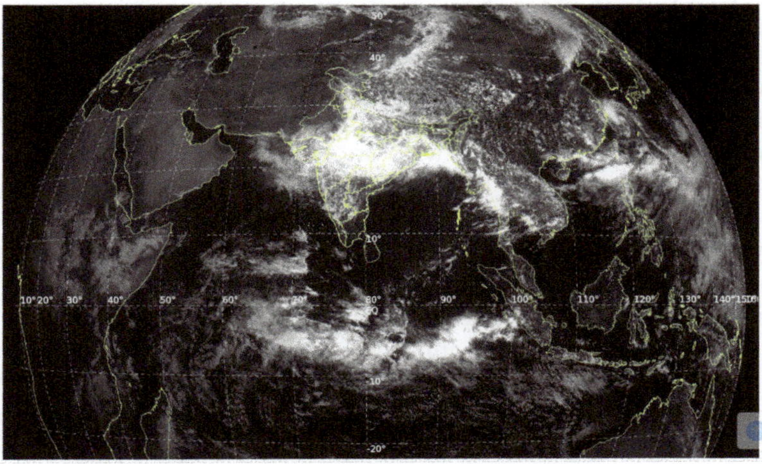

Figure 2.4.5.2 INSAT 3DS VIS full disc image for 29 July 2025 0530 UTC showing a prominent SHET. (Source: https://mausam.imd.gov.in/)

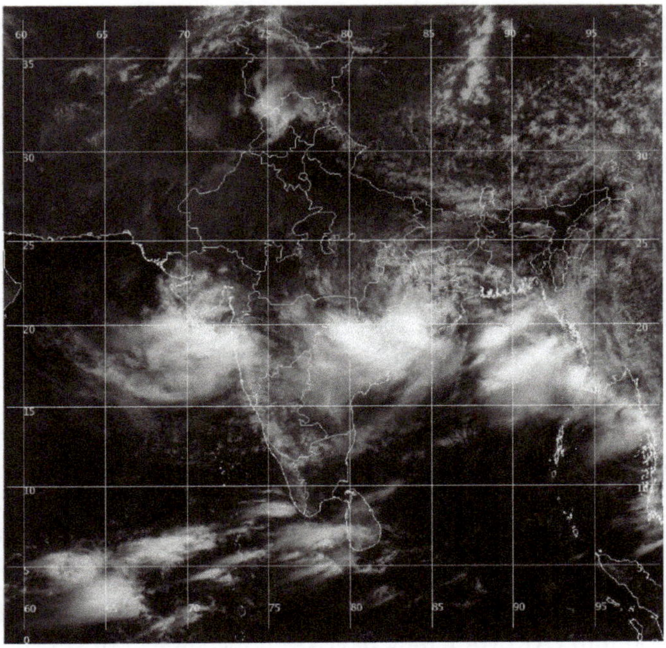

Figure 2.4.5.3 Kalpana-1 VIS image of 3 July 2006 showing a monsoon depression over eastern India flanked by intense convection on the Arabian Sea and Bay of Bengal and an off-shore trough along the west coast. The lower portion shows the cirrus blow-off due to the strong TEJ winds. (Source: IMD)

2.4.6 Breaks in the Monsoon

The southwest monsoon is in its most intense phase during July and August and these are the months of peak rainfall activity over India. However, the monsoon does not maintain a uniform strength all throughout this period and the wet spells are interspersed with dry spells over different parts of the country. At times, when the rainfall becomes very subdued over the country as a whole, the situation is described as a break in the monsoon. Breaks can be of varying intensity and duration, but they have certain common characteristic features. The most prominent synoptic development associated with a break is the shifting of the axis of the seasonal monsoon trough to the foothills of the Himalayas. The large-scale atmospheric flow pattern develops many anomalous features. The north-south pressure gradient along the west coast reduces from 9 to 3 hPa and lower tropospheric westerlies weaken along the west coast. On the other hand, westerlies over the Gangetic plains strengthen and their speeds may reach even 50 knots. These winds are dry and transport dust from Rajasthan into the Gangetic plains. The monsoon trough slopes northward instead of sloping southward.

In the case of shorter breaks in the monsoon, only the western end of the monsoon trough may shift to the foothills of the Himalayas, with the eastern end remaining entrenched in its normal position. In this case there is excess rainfall over western Himalayas and reduced rainfall over the plains of northwest India. Such a break may be induced by the passage of a western disturbance. The monsoon activity recovers after the western disturbance has moved away.

When the entire monsoon trough migrates to the Himalayan foothills, there is a prolonged break, which may last for a few days or even longer. There is excess rainfall all along the Himalayan foothills, especially on the eastern side, leading to a flooding of the Himalayan rivers, but rainfall is deficient over large parts of the country.

Breaks are most common in July and August. If a break-like situation develops in June, it is generally not called a break, as the monsoon may not have even established itself by then over some parts of the country. Likewise, a weakening of the monsoon in September may not be a break, but signal the beginning of the monsoon withdrawal process.

As the breaks in the monsoon strongly modulate the All-India Summer Monsoon Rainfall (AISMR), they have been the subject of several studies, in which different definitions of a break are also to be found (Webster et al 1998, De et al 2002, Gadgil et al 2003, Rajeevan et al 2010). The year 2002 witnessed a very prolonged break in July and the AISMR was deficient by 19 %. In the year 2005 there were long breaks during August and AISMR was deficient in that month by 28 %.

Clear advance synoptic indications of an impending break are not always available but satellite imagery is of great value in monitoring the migration of the monsoon trough from its normal position to the foothills and back. Satellite images during break situations have a conspicuous absence of clouds over a large part of the country.

Sometimes a break is caused by a monsoon low or depression moving from the head Bay towards Assam and the movement can be seen in satellite pictures. After a persistent break, its cessation may be induced by the formation of a monsoon depression in the Bay or the passage of a low pressure wave across south Indian peninsula that builds up the pressure gradient along the west coast. Here again satellite pictures provide valuable indications of the likely cessation of the break.

Jenamani (2004) identified two intermediate phases, one in which the monsoon trough is migrating from its normal position to the foothills and another in which the monsoon trough is returning from the foothills to its normal position and examined them with respect to rainfall, pressure and circulation features. He concluded that in the pre-break phase, there is large negative pressure anomaly and excess rainfall over north India and large positive pressure anomaly and deficient rainfall over peninsular India. In the post-break revival phase, the opposite conditions prevail. He also found that in the pre-break phase, monsoon disturbances form over extreme north Bay of Bengal and in the revival phase, they form over south Bay of Bengal. If these disturbances move northward or northwestward the monsoon revives, but if they move westward the break only gets prolonged and intensified.

2.4.7 Withdrawal of Southwest Monsoon

Although the period 1 June to 30 September is commonly regarded as the southwest monsoon season, it is more so because of the convenience of generating climatological statistics of rainfall and other atmospheric parameters than for sound physical reasons. While the onset of monsoon over Kerala occurs around 1 June, northwest India may not get its monsoon rains until the end of June. In effect, pre-monsoon rainfall in north and northwest India gets clubbed with the rainfall associated with the monsoon for climatological purposes. Likewise, the southwest monsoon starts losing its strength in early September over northwest India, where the monsoon season is effectively of only two months duration. On the other hand, over large parts of central and southern India, rains persist much beyond the date of 30 September.

The withdrawal of the southwest monsoon from northwest India is an event that is in no manner comparable to its onset over Kerala. While the onset over Kerala has been a subject of great scientific interest, there have been comparatively much fewer studies of the withdrawal of the monsoon. The withdrawal process is associated with the establishment of anticyclonic flow over northwest India, a reduction in the atmospheric moisture content and of course the cessation of rainfall. This is the beginning of the transition to the winter circulation pattern, and the return of the ITCZ to its southerly position.

Figure 2.4.7.1 Kalpana-1 WV image taken on 2 October 2015 showing the southwest monsoon in its withdrawal phase. (Source: IMD web site imd.gov.in)

There is no clear objective definition of the withdrawal, however, and there is no rule as to how long to wait before declaring that the monsoon rains have come to an end. There are many instances when the monsoon withdrawal has been followed by rains in north India as a result of the passage of a western disturbance or over peninsular India because of strong thunderstorm activity.

The Kalpana-1 water vapour channel image of 2 October 2015 (Figure 2.4.7.1) clearly shows the northern parts of India to be devoid of any moisture, whereas the peninsula continues to be cloudy and raining. Generally speaking, by 15 October, the southwest monsoon withdraws from the country except the southern peninsula, where the northeast monsoon continues to bring more rains as described in the next section.

2.4.8 Northeast Monsoon

Even after the withdrawal of the southwest monsoon, rainfall continues until mid-December over southern parts of India, and up to the first week of January over Sri Lanka. This rainfall is produced by the northeasterly winds in the lower troposphere associated with the northeast monsoon. Compared with the southwest monsoon, the northeast monsoon is a phenomenon of a much smaller scale, and it affects only the southern parts of the Indian peninsula during the months of October to December.

The mean date of onset of the northeast monsoon over Tamil Nadu is 20 October with a standard deviation of a week. However, there have been years of very early onset (5 October 1943) as well as extremely delayed onset (11 November 1915). The onset of the northeast monsoon is not rigidly defined, but it is known to be associated with several distinguishing features like the withdrawal of southwest monsoon up to coastal Andhra Pradesh, persistence of easterly/northeasterly winds up to 850 hPa in the lower troposphere and the occurrence of fairly widespread rainfall over coastal Tamil Nadu and adjoining areas. The onset is also marked by a reversal of the north-south pressure gradient with the lowest pressure being over the extreme southern peninsula, strengthening of upper tropospheric westerlies over north India and increase in moisture content over the southern peninsula.

The northeast monsoon gets intensified with the passage of easterly waves and low pressure areas across the Bay of Bengal (Figure 2.4.8.1). Tropical cyclones form at this time of the year, but as the northeast monsoon season advances, the cyclones move at relatively lower latitudes. In November and December, cyclones have a tendency to move towards Tamil Nadu and some systems may even emerge into the Arabian Sea.

For Tamil Nadu, northeast monsoon rainfall is more plentiful than southeast monsoon rainfall. The annual normal rainfall of Tamil Nadu is 100 cm, out of which 35 cm is received during the southwest monsoon and 48 cm in the northeast monsoon season. The northeast monsoon rainfall has a standard deviation of 14 cm making it highly variable. In a rare situation, on 1 December 2015, Chennai received a record-breaking rainfall of 35 cm, causing extensive flooding of the metropolis.

The northeast monsoon season ends in December. Again there is no specific definition of withdrawal of northeast monsoon. After December, the near-equatorial trough moves further south and the formation of tropical cyclones near the equator is inhibited.

Figure 2.4.8.1 INSAT-3D TIR1 image of 9 November 2015 showing an active northeast monsoon situation.
(Source: IMD web site imd.gov.in)

Khole et al (2003) have composited the area-weighted rainfall for the five meteorological sub-divisions of India forming the core region of the northeast monsoon and found that it is 11 % higher than the long period average in El Nino years and 6 % lower in La Nina years.

Raj (1996) has studied the thermodynamical aspects such as precipitable water vapour, upper air temperatures and moisture fluxes during different phases of the northeast monsoon. Suresh et al (2001) made a similar analysis based upon TOVS-derived outgoing longwave radiation (OLR) and precipitable water vapour for 1996-1998 over the northeast monsoon domain. They found that in the active phase of the northeast monsoon, the lower and upper atmosphere is warmer and the middle atmosphere cooler than in the weak phase. The OLR distribution suggests that during the active phase, the rainfall is maximum over south coastal Tamil Nadu and decreases towards the Bay of Bengal, while in the weak and post-withdrawal phases, there is higher rainfall over the Bay than over Tamil Nadu. Suresh et al (2002) have also made a study based on TOVS data of atmospheric boundary layer parameters during the northeast monsoon.

2.4.9 Easterly Waves

During the monsoon season, low pressure systems move from east to west across Myanmar into south and central Bay of Bengal and they help in the formation of monsoon depressions. During the pre-monsoon and post-monsoon seasons, some remnants of west Pacific Ocean typhoons travel across southeast Asia, enter into the Bay of Bengal and trigger the formation of tropical cyclones. Easterly waves can be easily tracked through satellite imagery (Figure 2.4.9.1)

Figure 2.4.9.1 Kalpana-1 IR image of 6 November 2015 showing an easterly wave over the Bay of Bengal.
(Source: IMD web site imd.gov.in)

Easterly waves are observed over other parts of the world also. These are waves of relatively feeble intensity. They have smaller amplitudes compared to troughs and ridges and cannot be easily identified on synoptic charts as usually the data is inadequate. They lack a uniform structure as they change with the nature of the zonal current which itself has varying regional and seasonal characteristics. In our region, easterly waves lie between the sub-tropical ridge and the ITCZ, both of which slope towards the equator with increasing height. Hence at any given latitude, there is considerable difference in the depth of the easterlies. In the Bay of Bengal, easterly waves give rise to closed lows which then move independent of the easterly waves themselves.

2.5 Tropical Cyclones

Getting caught in violent storms at sea has always been a common hazard associated with sea voyages. However, our earliest scientific information about tropical cyclones came from the distress signals sent out by ships and the experiences and observations recorded by mariners in their logs that were scanned by meteorologists when the ships berthed at the ports. It was only in the mid-twentieth century, that a cyclone could be caught in the view of a weather radar as it approached the coast, its structure examined and intensity estimated (Raghavan 1997). Afterwards, in the U.S., dare-devil aircraft reconnaissance missions were flown into tropical cyclones, enabling the *in situ* measurement of atmospheric parameters within and around them. After the launch of the world's first weather satellite in 1960, meteorologists got their first glimpse of a tropical cyclone from space. Initially, only one satellite image of a tropical cyclone was available in a day, and night-time observations became possible later. Early satellite imagery, though quite primitive by present standards, provided a new insight into the structure of tropical cyclones (Fett 1964) and ushered in a new era of tropical cyclone observations and forecasting.

Subsequently, with the launch of geostationary satellites like GOES, Meteosat and GMS, and that of the first INSAT in 1982, tropical cyclones came under a global round-the-clock surveillance right from their genesis up to landfall. Satellite imagery is most useful at all stages of development of tropical cyclones, for first detecting their formation far out at sea, then following their movement, determining their intensity and observing changes in their characteristics.

It has been known for long that there are three basic pre-requisites for the formation of a tropical cyclone, or cyclogenesis, over the tropical oceans. These are:

(a) the Coriolis force which produces the rotation effect,

(b) a warm ocean that provides energy to the cyclone, and

(c) relatively moist layers in the middle troposphere.

Because of the requirement of the Coriolis force to provide for near-gradient wind balance to occur, a minimum distance of about 500 km away from the equator is required. Without the Coriolis force, the low pressure of the disturbance cannot be maintained. The sea surface temperature should be least 26.5 °C, and temperature should be high enough to a sufficient depth of the order of 50 m below the surface. The atmosphere should cool rapidly with height so that it is potentially unstable to moist convection. It is the thunderstorm activity which allows the heat stored in the ocean waters to be released for the tropical cyclone development. Dry mid-tropospheric levels are not conducive for allowing the continuing development of widespread thunderstorm activity.

Besides the three conditions mentioned above, another condition for cyclogenesis is a pre-existing low-level disturbance. Tropical cyclones do not form on their own, and they require at least some weakly organized system at the surface with sufficient vorticity and low level inflow or convergence. What is also required is a weak vertical wind shear between the surface and the upper troposphere. Large values of vertical wind shear work contrary to cyclogenesis. If a tropical cyclone has already formed, large vertical shear can weaken or destroy the tropical cyclone by interfering with the organization of deep convection around the cyclone centre.

While all these conditions are necessary, they are not always sufficient. Many disturbances that seem to be in the midst of a favourable environment, do not develop into tropical cyclones. Also, all cyclones do not have devastating effects. Those which form close to the coastline and have no scope for intensification may even turn out to beneficial as they may give just the badly needed precipitation over the coastal areas. Some systems may move into colder waters and dissipate there without making a landfall. Then there are storms which recurve in the middle of a long path and move away from the coast, sometimes even after having almost reached it. As a result of all these possible situations, full-blown killer storms are, thankfully, not that common.

A tropical cyclone transports large amounts of sensible and latent heat energy from the sea surface to the atmosphere. As the cyclone moves away, SST falls by several degrees due to upwelling of colder waters from deeper layers, vertical mixing, and cooling due to evaporation at the sea surface because of strong winds. While high SSTs are favourable for cyclogenesis, a tropical cyclone can itself cool down the sea in its wake. This acts as a partial brake on the intensification of cyclones. Hence, one tropical cyclone is not immediately followed by another, and at least a few days must elapse in between.

2.5.1 Nomenclature of Low Pressure Systems

Generally speaking, a tropical weather system that crosses the 64 knots or 118 km/hr wind speed threshold gets identified as a major storm. However, different meteorological agencies responsible for tropical cyclone forecasting use different nomenclature practices to describe the evolution of the system through successive stages. Table 2.5.1.1 shows how the India Meteorological Department (IMD) classifies cyclonic systems over the Arabian Sea and the Bay of Bengal. It uses the term "Very Severe Cyclonic Storm" to categorize a tropical cyclone with wind speeds exceeding 64 knots. In the eastern and central Pacific Ocean and north Atlantic Ocean, such a system is called a "Hurricane" and in the northwest Pacific Ocean it would be termed as a "Typhoon". These differences, however, are only in the nomenclature and not in the characteristics of the storm, and it is at this stage that it gets a name identifier like Katrina or Rita as per naming conventions.

Table 2.5.1.1 Nomenclature of Tropical Weather Systems used in India.

System Category	Wind Speed (knots)	Wind Speed (km/hr)
Low Pressure Area	16 or lower	30 or lower
Depression	17-27	31-49
Deep Depression	28-33	50-61
Cyclonic Storm	34-47	62-88
Severe Cyclonic Storm	48-63	89-117
Very Severe Cyclonic Storm	64-90	118-167
Extremely Severe Cyclonic Storm	91-119	168-221
Super-Cyclone	120 or higher	222 or higher

In the U.S., hurricanes are further ranked on the 5-point Saffir-Simpson scale ranging from Category 1 which has the lowest damage potential, to Category 5 which denotes the most powerful hurricane. In India, the term "Super-Cyclone" is used for cyclones in which the winds are 120 knots or 222 km/hr or higher, which roughly corresponds to a Category 4 hurricane on the Saffir-Simpson scale.

2.5.2 Tropical Cyclone Basins

On an average, about 85 tropical cyclones form every year over seven major oceanic basins around the world. These are of tropical storm intensity with wind speeds exceeding 63 km/h or 34 knots. Of these, about 47 reach hurricane/typhoon strength with wind speeds exceeding 118 km/h or 64 knots, and 20 become intense tropical cyclones of at least Category 3 intensity. Tropical cyclone activity, however, varies significantly across basins (Table 2.5.2.1).

Table 2.5.2.1 Formation of Tropical Cyclones over Different Ocean Basins.

Basin	Area	Average Number of Storms per Year	Percentage of Annual Global Total Number	Main Storm Season	Period of Peak Activity
Western North Pacific Ocean	From the dateline to Asia including the South China Sea	26	31	July-November, but storms occur throughout the year	Late August/early September
Eastern North Pacific Ocean	From Mexico to about the dateline	14	16	May-October	Late August/early September
Southwest Pacific Ocean	From 142 °E to about 120 °W	9	11	October-May	February-March
North Atlantic Ocean	Including the Gulf of Mexico, and the Caribbean Sea	14	16	June-November	September
North Indian Ocean	Including the Bay of Bengal and the Arabian Sea	6	7	April-June and October-December	Double peak in May and November
Southeast Indian Ocean	From 100 °E to 142 °E	7	8	October-May	January-March
Southwest Indian Ocean	From Africa to about 100 °E	9	11	October-May	January-March

Over the south Atlantic Ocean, cyclones are the rarest, with the solitary exception of the one that hit Brazil in March 2004. Over the western north Pacific Ocean, cyclones are the most frequent, their number being as high as 30 % of the storms occurring annually over the entire globe. The Arabian Sea and Bay of Bengal get an average of only 4 to 6 cyclones per year, but historically some of them have been the world's deadliest. They are most frequent in the months of April-May (summer season) and October-November (post-monsoon season). Arabian Sea storms are comparatively fewer than Bay of Bengal storms.

The World Meteorological Organization (WMO) has designated six Tropical Cyclone Regional Specialized Meteorological Centres (RSMCs) and six Tropical Cyclone Warning Centres (TCWCs) with responsibility for different basins. Together they provide advisories and bulletins with up-to-date first level basic meteorological information on all tropical cyclones, hurricanes or typhoons all over the world. The RSMCs are in fact cooperative ventures operated by national meteorological agencies and their locations and areas of responsibility are as follows:

1) RSMC Miami-NOAA National Hurricane Center: Caribbean Sea, Gulf of Mexico, North Atlantic and Eastern North Pacific Oceans
2) RSMC Tokyo-JMA Typhoon Center: Western North Pacific Ocean and South China Sea
3) RSMC New Delhi-India Meteorological Department: North Indian Ocean
4) RSMC La Réunion-Météo-France Tropical Cyclone Centre: Southwest Indian Ocean
5) RSMC Nadi-Fiji Meteorological Service Tropical Cyclone Centre: Southwest Pacific Ocean
6) RSMC Honolulu-NOAA Hurricane Center: Central North Pacific Ocean

2.5.3 Structure of Tropical Cyclones

The life cycle of a tropical cyclone consists of four phases, the formative phase, the developing phase, the mature phase and the dissipating phase.

(a) *Formative Phase*: In the neighbourhood of the ITCZ, relative cyclonic vorticity is already present. An easterly wave helps in the formation of a weak 'cold core' low in which there is relatively colder air at the lower levels and relatively warmer air at the higher levels. Cyclonic vorticity induces horizontal inflow in the frictional layer and makes the air to rise dry adiabatically. Satellite pictures show spiraling clouds interspersed with open spaces.

(b) *Developing Phase*: The rising air, after reaching the condensation level, is saturated and warmer than the environment. Release of latent heat of condensation is a source of heat. This releases the convective instability of the lower and middle troposphere. Vertical acceleration of air induces inward spiralling of air. The warm core descends down to the sea level and the entire vortex becomes a warm core system.

(c) *Mature phase*: The winds strengthen, pressure falls and there is heavy rain at the surface. Descending motion starts at the top of the warm core and penetrates gradually downwards. This leads to the formation of the eye. The cyclonic vortex drifts in the general air current but also interacts with it and even modifies it.

(d) *Dissipating Phase*: After landfall, being deprived of warm moist air and encountering rough terrain, it again becomes a cold core system. It weakens into a depression and dissipates gradually. In rare cases, a cyclone which has dissipated over land may once again emerge into the ocean. For example a Bay of Bengal storm may weaken after crossing into the Indian peninsula, but continue to move westward and eventually emerge over the Arabian Sea to re-intensify. Some storms may move to higher latitudes and take on the characteristics of an extra-tropical cyclone. The total life span of a tropical cyclone may range from 5 to 15 days.

Figure 2.5.3.1 SuomiNPP satellite VIIRS image of 17 May 2021 showing extremely severe tropical cyclone Tauktae over the Arabian Sea prior to landfall on the Gujarat coast.
(Source: https://modis.gsfc.nasa.gov/gallery/)

A typical satellite view of a fully mature tropical cyclone is shown in Figure 2.5.3.1. The characteristic features are explained below:

(a) *Eye and eyewall*: At the centre of the pattern is the eye, which is a roughly circular area characterized by fair weather, comparatively light winds, lowest surface pressure and warmest temperatures aloft. Eyes range in size from 8 to 200 km, but most commonly have a 30-60 km diameter. The eye is surrounded by the eyewall which is a ring of deep convection. It is the area of highest surface winds in the tropical cyclone. The eye is composed of air that is slowly sinking while the eyewall has a net upward flow as a result of many strong updrafts and downdrafts. The high temperatures of the eye are due to compressional warming of the subsiding air. The appearance of an eye is indicative of the strengthening of the storm. It increases the dynamical stability. As the storm intensifies, the eye shrinks in size and the eye area becomes more cloudfree. In very intense systems, there may be two cloudfree regions surrounded by two cloud walls, resulting in a double eyewall structure. The eye is mostly round, but occasionally eyes have been seen to have polygonal shapes as a result of instability of the wind structure near the radius of maximum winds. Polygonal eyewalls do not always have clear sky and calm winds but they can have small vortices associated with tall clouds and high winds.

(b) *Spiral bands*: Convection in tropical cyclones is organized into long, narrow rain bands which are oriented in the same direction as the horizontal wind. Because these bands seem to spiral into the centre of a tropical cyclone, they are also called spiral bands. Along these bands, low level convergence and upper level divergence are most pronounced. The width of the rain bands is about 100 km near the outer periphery of the storm but it reduces inwards. Severe convective rain rates of 3 cm/hr or more are associated with the rain bands. They move along with the system but also move around the centre in the direction of the tangential wind of the cyclone but much slower. Spiral bands persist for a few days but individual cells within the bands are very short-lived.

(c) *Central dense overcast*: Between the eyewall and the outer bands is the central dense overcast or CDO, which is a shield of cirrus clouds. This cirrus cloud shield results from the thunderstorms in the eyewall of a tropical cyclone and its rain bands. Before the storm reaches hurricane strength of 64 knots the CDO looks uniform. After that an eye may be seen in either the IR or VIS channel images. Circular CDOs are indicative of favourable, low vertical shear environments.

(d) *Moat*: This term usually refers to the region between the eyewall and an outer rain band and signifies a region of relatively light rain.

2.5.4 Dvorak Technique

The advent of meteorological satellites in the 1960's made an immediate impact on the detection and monitoring of tropical cyclones and their track prediction (Sikka 1971a, 1971b, Kulshrestha 1971). Soon thereafter, the technique formulated by Dvorak (1975) for intensity estimation of tropical cyclones from VIS satellite imagery available once a day gained widespread acceptance. The scope of the technique was later enlarged to include IR imagery which became available at night-time also (Dvorak 1984). Dvorak technique is basically a pattern recognition process which assumes that certain characteristics or features of the cloud organization are indicators of the intensity. The current satellite picture of a tropical cyclone has to be matched against a number of possible pattern types or analogues. The features, however, may not always be clearly defined or they may appear differently or be uncharacteristic of the standard patterns. Such a complexity can be dealt with through a set of systematic procedures, quantifiable and measurable cloud pattern descriptions and models of storm development with time, which are stipulated in Dvorak technique.

Table 2.5.4.1 Intensity of Tropical Storms as per Dvorak Technique.

CI-Number or T-Number	Maximum Sustained Wind (knots)	Central Pressure (hPa)	
		Atlantic Ocean	NW Pacific Ocean
0.5	25	----	----
1.0	25	----	----
1.5	25	----	----
2.0	30	1009	1000
2.5	35	1005	997
3.0	45	1000	991
3.5	55	994	984
4.0	65	987	976
4.5	77	979	966
5.0	90	970	954
5.5	102	960	941
6.0	115	948	927
6.5	127	935	914
7.0	140	921	898
7.5	155	906	879
8.0	170	890	858

The application of Dvorak technique has been discussed in detail by Kalsi (2006) in a case study of the Orissa super-cyclone of 29 October 1999. Dvorak's model of cyclone development is built in terms of what is called the T-Number (T = Tropical). This has a range of values of 1 to 8 is steps of 0.5, associated with increasing wind speed and decreasing sea level pressure (Table 2.5.4.1). T1 stands for a nascent disturbance, T2 for a pre-storm situation and T2.5 for a marginal tropical cyclone. T3.5 is a severe cyclonic storm, T4 - T6 a very severe cyclonic storm and T6.5 or higher a super-cyclone. T8 would describe the most extreme case. The T-Number would normally increase by 1 per day during intensification but departures from the model rate are possible.

Figure 2.5.4.1 MODIS image of 7 August 2006 showing three different typhoons over the western Pacific Ocean: Saomai, Maria and Bopha with strengths in decreasing order.
(Source: NASA web site earthobservatory.nasa.gov/)

Figure 2.5.4.1 shows a satellite image of three tropical storm systems co-existing in close proximity over the western Pacific Ocean, but at different stages of evolution, each pattern corresponding to different T-numbers on Dvorak's scale. Dvorak technique of tropical cyclone intensity estimation has the nature of a complex decision tree with several branches. Since it was intended for operational use by duty forecasters working in meteorological centres, it was the form of a detailed rule-book. Only the 10 basic steps of the decision making process are described below:

Step 1 - Locating the Cloud System Centre (CSC): The first step is to locate the CSC which is the focal point of all curved lines or bands of the cloud system, or it is the geometric centre of the eye if present. A disturbance showing signs of reaching tropical cyclone intensity is classified as T1 if it has persisted for 12 hours or more, if it has a CSC defined within an area having a diameter of 2.5° lat or less, which has persisted for 6 hours or if it has an area of dense, cold (<–31° C) overcast greater than 1.5° lat in extent that appears less than 2° lat from the centre. The overcast may also appear in cumulonimbus lines that curve around the centre. A considerable variability of cloud patterns is often observed during the T1 stage. These weak systems do not involve deep tropospheric circulations. They are short-lived and prone to diurnal and non-periodic oscillations in convection. Kalsi (1993) examined several time sequences of INSAT-1B images and observed that initially there may not be a perceptible difference between developing and non-developing mesoscale cloud clusters over the ocean, but the persistence of convection as a prelude to cyclogenesis is a feature common to many Bay of Bengal tropical cyclones.

Step 2 - Determining the Pattern Type: The manner in which the CSC is defined determines which of the following pattern types would be used: (2a) Curved Band Pattern, (2b) Shear Pattern, (2c) Eye Pattern, (2d) Central Dense Overcast or CDO Pattern, (2e) Embedded Centre Pattern or (3) Central Cold Cover Pattern. When the cloud pattern being analyzed does not conform to any of these patterns, Step 3 is used.

Step 2a - Curved Band Pattern: This pattern is very commonly observed in both VIS and IR satellite imagery in the development stage of a tropical cyclone (Typhoon Bopha in Figure 2.5.4.1). The intensity estimate is derived from the extent to which the cloud band encircles the storm centre, which is measured as the arc length of the curved band fitted to a 10° logarithmic spiral overlay. The spiral overlay is fitted to the axis of the coldest gray shade (most dense clouds) within the cloud band. Depending on the fit, the associated Data T-Number (DT) is obtained from the chart, either VIS or IR.

During the first two days of development (T1 to T2), the band curvature may increase, remain unchanged or even decrease for some time. So, on the first day, the T-Number is increased from T1 to T2 unless there are clear signs of weakening. It is also important to allow at least 24 hr to pass between a T2 and a T3 classification as even though the coiling process may be rapid, the surface pressure may not fall accordingly.

Step 2b - Shear Pattern: The curved cloud bands may sometimes get obliterated due to strong vertical wind shear. In such a case, the dense upper level clouds get separated from the low level circulation. The dense cloud mass appears to be distant from the centre defined by the cumulus lines (Figure 2.5.4.2). Shear patterns most commonly occur during early development and weakening. They are identified by the cold clouds moving to one side of the cyclone and developing a sharp edge. The DT is derived from both the method of defining the CSC and the distance between the low cloud centre and the dense, cold overcast.

Figure 2.5.4.2 Kalpana-1 IR image of 7 May 2004 showing shear pattern in an Arabian Sea tropical cyclone. (Source: IMD)

For tropical storm intensity (T2.5 - T3.5), the centre will be defined by parallel, circularly curved low cloud lines with a diameter of 1.5° lat or less near or under the edge of a dense, cold (<–31° C) overcast cloud mass. For weaker systems (T1.5 - T2.5), the low cloud centre will either be poorly defined in spiral lines within 1.25° lat of the cold overcast clouds, or circularly defined near a small, dense overcast (< 1.5° lat diameter).

Step 2c - Eye Pattern: At T2.5, the cloud band may curve half-way around the centre. As the intensity increases, the band widens and coils to form a gap or hole in the centre which is called a banded eye (Figure 2.5.4.3). When the cyclone develops an eye embedded in the CDO, the intensity can be estimated from the size, shape and appearance of the eye and the smoothness of the CDO (Figure 2.5.4.4). Eye patterns are analyzed for systems which have been at T2 or greater for 24 hours and have an eye. The DT is derived from the Eye Number E, Eye Adjustment Number EA and Banding Feature Number BF for which there are separate rules.

Figure 2.5.4.3 Kalpana-1 VIS image of 28 April 2006 showing a banded eye in a Bay of Bengal tropical cyclone. (Source: IMD)

Step 2d - CDO Pattern (VIS imagery): CDO patterns consist of a dense, solid-looking mass of clouds covering the cloud system centre (Figure 2.5.4.5). Generally, well-defined CDOs of at least 1° lat width are associated with tropical storm intensities and those measuring 2° lat or more are associated with hurricanes. Banding features (BF) and central features (CF) terms are added to get the DT.

Figure 2.5.4.4 GOES-West VIS image of 6 September 2025 2200 UTC showing hurricane Kiko over north Pacific Ocean with a well developed eye. (Source: https://tropic.ssec.wisc.edu/)

Figure 2.5.4.5 Kalpana-1 IR image of 2 October 2004 showing CDO pattern in an Arabian Sea cyclone. (Source: IMD)

Step 2e - Embedded Centre Pattern (IR imagery): Embedded centre patterns are analyzed when the tropical cyclone has been T3.5 or stronger and when the CSC is clearly indicated to be within a cold overcast (<9° C). Curved cloud lines or bands within the cold overcast as well as the outer curved bands will indicate the location of the CSC within the overcast. The analysis of this pattern is similar to the eye pattern analysis except that no eye adjustment factor is added.

The non-availability of VIS images at night presents many additional difficulties in the operational use of the Dvorak technique. Firstly, IR sensors generally have a poorer resolution than VIS sensors. Secondly, thin cirrus clouds which are transparent to VIS radiation, show up in IR pictures prominently, blocking the view of underlying cloud features. Thirdly, the low cloud lines which are clearly seen in VIS pictures and help to locate the centre are not noticeable in IR images. However, the digital brightness data, if used directly, can be of great value in quantitative analysis and digital enhancement procedures can be applied to bring out many of the storm features. Kalsi (2006) had applied the digital gray scale enhancement technique to the sequence of images of the Orissa super-cyclone of 29 October 1999 and brought many additional features which cannot be seen in VIS and un-enhanced IR images. The details of the enhancement scheme have been given later in this chapter in Section 2.5.6.

Step 3 - Central Cold Cover (CCC) Pattern: The CCC pattern consists of a near-circular, cold or dense overcast covering the cyclone centre or comma head and obscuring the expected signs of pattern evolution. It is usually associated with arrested development and is not to be confused with a very cold comma pattern, which has smooth texture and an indication of a wedge and has curved cirrus lines around it.

Step 4 - Past 24-hour Trend: The trend of the past 24-hr intensity change is determined by other observations, or by comparing the current and 24-hr past cloud features. Development is generally associated with increased organization and better defined central features.

The signs of development are: (D1) curved band coils tighter around the CSC, (D2) CDO becomes larger or an increase in banding features is noted, (D3) CSC becomes more tightly defined in curved cloud lines or appears closer to the dense overcast, (D4) eye is more embedded, more distinct (warmer), less ragged, or is surrounded by colder (smoother textures) clouds, or more banding features, and (D5) no significant (non-diurnal) warming of the cloud system is noted. The signs of weakening (W) are exactly the opposite of signs D1 - D5.

Signs of steady state are: (W1) a central cold cover appears in a T3.5 or stronger tropical cyclone or has persisted for more than 12 hr in a weaker cyclone, (W2) the CSC's relationship to the cold clouds has not changed significantly, and (W3) there are conflicting indications of both development and weakening.

Step 5 - Model Expected T-Number (MET): MET is determined by using the 24-hr old T-Number, the D, S, or W, decision in Step 4, and the past amount of intensity change of the cyclone. When the growth rate has not been established in the case of new developments or reversals in trend, a past rate of change of one T-Number per day is assumed. Rapid or slow past rates of change are established when two consecutive analyses showing rapid or slow pattern evolution are observed at 6 hr or longer intervals, or when one observation of strong intensification or weakening is made.

Step 6 - Pattern T-Number (PT): The pattern T-Number is used primarily as an adjustment to the MET.

Steps 7 and 8 - Final T-Number: The final T-Number is determined from the rules stipulated in the technique. Constraints are applied to hold the T-Number within acceptable bounds and avoid misinterpretations of marked changes.

Step 9 - The Current Intensity (CI-Number): The CI-Number relates directly to the intensity of the tropical cyclone and is determined from the T-Number by the rules. After each intensity analysis, the previous analysis of the tropical cyclone has to be reviewed for consistency. If the previous day's analysis was erroneous, the T-number has to be corrected to provide a more accurate model-expected intensity. This correction may at times alter the current intensity analysis.

When the storm begins to weaken, the CI-Number is used to indicate the storm intensity in place of the T-Number. The CI-Number is held higher than the T-Number for weakening storms, otherwise they are the same. This is done as a precaution as indications of weakening may sometimes only be temporary.

Step 10 - Forecasting the intensity: Forecasts of change in cyclone intensity are made using the Dvorak model development curve, statistical methods, and subjective assessment of satellite imagery and environmental conditions. It is necessary to consider whether the cyclone is entering or leaving an environment of strong vertical wind shear or approaching land.

It is important to note here that tropical cyclones show a lot of diurnal variation and the cloud patterns change accordingly from morning to evening. Heavy rain follows the same diurnal cycle as convection. The maximum occurs between local midnight and dawn over the ocean, but there is no preferred time over land, especially when orographic effects dominate.

2.5.5 Objective Dvorak Technique

Extensive use of Dvorak technique brought to light many of its limitations and subjectivity. Dvorak technique was initially calibrated with reference to aircraft observations over the Atlantic and Pacific Oceans and basin to basin differences in storm evolution and structure were not accounted for. New types of satellite data in microwave and WV channels later became available operationally and they provided additional information to cyclone forecasters in making a judicious use of Dvorak technique (Velden et al 1991).

The Objective Dvorak Technique (ODT) was one of the first attempts to create an objective, automated version of the technique (Velden et al 1998). It aimed to improve upon Dvorak's original subjective rules and logic using computer algorithms. ODT used digital IR data but still required some manual input, such as the user having to select the storm center location. Its use was often limited to stronger cyclones of hurricane/typhoon strength and it could not be applied to the entire lifecycle of a storm or to weaker systems.

ODT extracts two temperatures from the IR data in order to obtain an intensity estimate. It first determines if the storm has an eye. If so, it uses the warmest pixel temperature within a 40 km radius of the chosen storm centre. This value is retained as the 'eye temperature'. In the absence of an eye, the value of the pixel at the user-defined storm centre location is used as the eye temperature. Next, ODT analyses temperatures on concentric rings centred on the eye to determine the surrounding temperature. This parameter is a proxy for the vigour of the eyewall convection which is associated with storm intensity. The intensity is then determined by matching them with standard cloud patterns objectively by examining areal histograms of cloud top temperatures and corresponding Fourier Analysis for the eye region and surrounding cloud region.

ODT was subsequently further refined and named as Advanced Dvorak Technique (ADT) which is also objective but makes use of microwave and WV data as well. ADT is fully automated and completely removes human subjectivity from the intensity estimation process. It uses objective algorithms to determine the storm centre and cloud patterns. It is applicable to all phases of a tropical cyclone's life cycle, and to weaker storms and subtropical systems (Olander et al 2019).

However, Dvorak's original technique continues to be the standard method for estimating TC intensity over all tropical regions outside the North Atlantic and Caribbean Sea where aircraft reconnaissance is not available in spite of its limitations. Ahmed, Mohapatra and others (2021a, 2021b) have made a new evaluation of the Advanced Dvorak Technique over the north Indian Ocean, in the particular context of supercyclone Amphan of May 2020 as observed through satellite images.

2.5.6 Image Enhancement

In satellite images, the brightness of a pixel is usually quantified on a scale ranging from 0 to 255, or from 0 to 1023, which is called the gray scale. The number 0 on the gray scale stands for pure black and the number 255 or 1023 stands for pure white. This scale is shown as a strip of continuously increasing brightness at the top or bottom of a satellite image for reference. In a VIS image, lower values on the gray scale are seen over oceans because of their high absorption of solar radiation. The higher gray scale values are seen over snow covered regions and Cb cloud tops. In an IR image, the lower end of the gray scale corresponds to high temperatures and the higher end of the gray scale to low temperatures. In a WV image, lower values of the gray scale correspond to dry regions and higher values to moist regions.

While the gray scale is basically linear, in order to improve the contrast between individual pixels, a non-linear gray scale may at times be applied to the image. This procedure is called enhancement, as it effectively enhances the original gray scale in some parts and brings into prominence what may otherwise be overlooked. Of course, stretching the gray scale in one part implies that it will get compressed in another part. Therefore, enhancement of any feature is always at the cost of some other feature. Various enhancement curves can be designed and used for meeting different requirements. For example, in Dvorak technique a what is known as a BD curve is applied to tropical cyclone images to enhance the outer cloud features and separate the class boundaries. Here, B stands for Black-and-white and D for Dvorak (Figure 2.5.6.1).

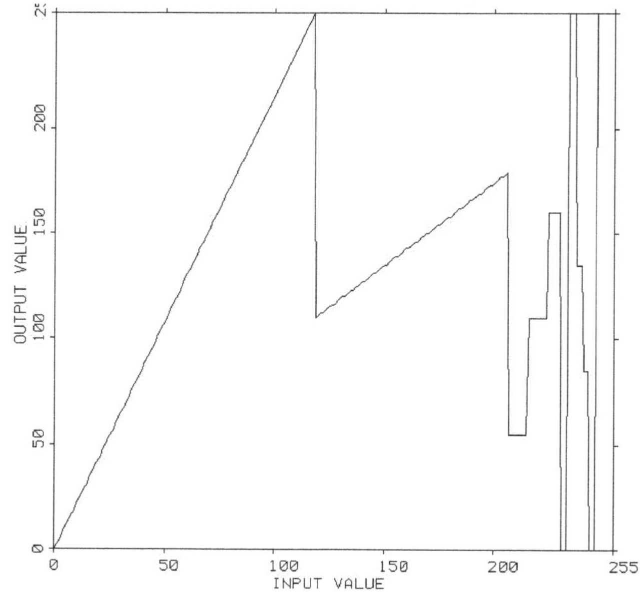

Figure 2.5.6.1 Gray scale enhancement BD curve for tropical cyclone images (Source: IMD)

Pseudocolour imagery is produced by another kind of enhancement in which parts of the gray scale are not stretched but assigned different colours or shades to bring out the essential features. For example, tall cloud tops can be made to appear as red or clear ocean as blue. In pseudocolour images, the possible enhancement choices are much greater and can be effectively used to draw attention to the desired features. Another way of generating false colour images is by combining VIS and IR images. Here as the common features get added and others get subtracted, the combined colour image can be made to highlight the essential features or give a natural look. This is called the RGB enhancement method. Mitra et al (2018, 2019) have described in detail how different RGB combinations for day and night are applied to INSAT-3D images to distinguish between fog, snow and low clouds.

Erin was the first major hurricane of the 2025 Atlantic season. It began as a tropical depression in the north Atlantic Ocean west of the Cabo Verde Islands. It quickly intensified on 15 August 2025 into a Category 5 hurricane with winds nearing 250 km/hr, making it one of the fastest-strengthening Atlantic hurricanes on record. It weakened thereafter and curved northeast between the U.S. East Coast and Bermuda, avoiding a direct landfall. Its rapid intensification is linked to unusually warm Atlantic waters and low wind shear, conditions increasingly common due to climate change. Figures 2.5.6.2 to 2.5.6.4 show the VIS-SWIR channel image on 17 August, its BD enhancement and pseudocolour enhancement respectively.

Figure 2.5.6.2 VIS-SWIR image of Hurricane Erin on 17 August 2025 2330 UTC. (Source: https://tropic.ssec.wisc.edu/tropic.php)

Figure 2.5.6.3 BD curve enhanced image of Hurricane Erin on 17 August 2025 2330 UTC. (Source: https://tropic.ssec.wisc.edu/tropic.php)

Figure 2.5.6.4 Pseudocolour image of Hurricane Erin on 17 August 2025 2330 UTC. (Source: https://tropic.ssec.wisc.edu/tropic.php)

2.5.7 Movement of Tropical Cyclones

The real impact of tropical cyclones depends on several factors. While the scale of destruction of property depends upon the intensity of the system and the associated wind speeds, rainfall amounts and rain rates, the loss of human life is more often caused by the storm surge and inundation. While the winds and rainfall are related to the degree of organization of the cyclonic system, the storm surge depends to a great extent on the nature of the coastline at the place of landfall, the slope of the sea bed and the land use pattern. That is why accurate track prediction several days in advance of expected landfall is important.

While a tropical cyclone is characterized by strong winds of the order of 100 to 200 km/hr, the cloud system as a whole moves only at about 20 km/hr along with the large-scale atmospheric flow around it. Usually the system moves towards the west, but has some northward movement as well, because of the existence of an east-west axis of high pressure called the sub-tropical ridge to the north of the system. If the subtropical ridge is weak, the system turns northward rather than westward, and as a result, away from the coastline and ultimately, it may turn fully eastward. This behaviour is called recurvature. The problem is to predict if the storm will recurve or go straight ahead and make landfall. Recurvature is the cause of many tropical cyclones over the Bay of Bengal first heading towards Andhra Pradesh, then turning towards Orissa and finally making landfall over Bangladesh or Myanmar. Occasionally, a cyclone may even perform a loop in its path.

The movement of a tropical cyclone can be predicted in two simple ways. One is by reference to the climatology of storms for that region and time of the year. If it is assumed that a cyclone will follow the same track as the past storms, a "climatological" forecast can be made for its direction and speed of movement. The other way is to view an animated sequence of successive satellite images over the previous 24 hours and assume that the cyclone will continue to move in the same direction and average speed. This is called a "persistence" forecast. The so-called CLIPER technique employs a combination of climatology and persistence.

Because of the wealth of data that the new generation of satellites are making available, the ability of meteorologists to make predictions of cyclone tracks has been consistently improving. In order to forecast what path a tropical cyclone will traverse over the ocean, it is necessary to first predict the steering current in which the tropical cyclone is embedded, and how the cyclone will interact with it. Numerical weather prediction models require appropriate data on the initial conditions of the atmosphere. However, even the most advanced models are constrained by the availability of initial data about the cyclone's internal structure and its external environment. The accuracy of cyclone track prediction by operational numerical weather prediction centres has steadily and surely increased over the years, making it possible to undertake disaster mitigation measures like evacuation of people much in advance.

Kelkar (1997) has discussed various satellite-based methods developed or attempted for cyclone track prediction. Velden et al (1984) used the satellite-derived mean layer wind as the steering current for forecasting hurricane movement. In spite of the difficulties in deriving cloud motion winds over cloudy regions, by combining them with microwave data, a fair approximation of the steering current can be derived.

Changes in upper level moisture patterns as seen from WV satellite imagery have also been used to forecast cyclone motion (Dvorak et al 1992). Two features of WV imagery are indicators of recurvature of northeast Pacific hurricanes. One is a curved moisture band about 10° northeast of the hurricane, and bowed cyclonically towards it. Another is an increasing north-south orientation of the moisture cloud pattern of the system. The WV imagery also helps in forecasting a westward turn in the track of the cyclone as it shows up a drying or darkening of the moisture pattern to the north prior to the turn.

The India Meteorological Department (IMD) uses a sophisticated blend of observational data, numerical models, and ensemble techniques to predict cyclone tracks over the Indian seas (Mohapatra et al 2025). It uses data from a network of buoys and AWS stations, and radar inputs when the system comes close to the coast. It runs a variety of NWP models and uses the Multi-Model Ensemble (MME) technique for track prediction. In recent years, IMD's track predictions have been near perfect, resulting in a significant reduction in the loss of life and damage.

2.5.8 Cyclone Monica in the Southern Hemisphere

Much fewer tropical cyclones develop in the southern Indian and Pacific oceans compared to the ocean basins in the northern tropics (Table 2.5.2). The South Atlantic Ocean basin does not even find a place in global statistics of tropical cyclones. On 28 March 2004, a rare tropical cyclone in the South Atlantic Ocean crossed the coast of Santa Catarina and Rio Grande do Sul, in the extreme southern parts of Brazil. This was a Category 1 hurricane with central winds of 125 km/hr, making it the one and only hurricane in the South Atlantic in recorded history.

Southern hemisphere cyclones have not been studied with as much interest as northern hemispheric cyclones, probably because many of them dissipate over the ocean and do not cause great destruction over land. Structurally speaking, tropical cyclones of both northern and southern hemispheres are in no ways different, but the characteristic feature so obvious in satellite images is the clockwise rotation of southern hemispheric cyclones compared with the anticlockwise rotation of northern hemisphere cyclones (Figure 2.5.8.1). This is the effect of the Coriolis force, which is associated with the rotation of the earth and causes the wind to deflect to the right in the northern hemisphere and to the left in the southern hemisphere.

Tropical Cyclone Monica was one of the most intense tropical cyclones ever recorded in the Australian region. It originated from a low pressure area near Papua New Guinea on 16 April 2006, off the northeastern coast of Australia. It first hit Far North Queensland near Lockhart River on 19 April as a Category 3 storm. It crossed Cape York Peninsula with weak winds, but on 20 April it reorganized itself and later-intensified over the Gulf of Carpentaria. Monica made a second landfall near Maningrida, Northern Territory on 24 April as a Category 5 storm. It grazed across the top of the Northern Territory of Australia, threatening the city of Darwin with heavy rains and very high winds.

The MODIS image acquired by the Aqua satellite on 24 April 2006 (Figure 2.5.8.1) shows Cyclone Monica as an impressively large and powerful storm with a compact eye and intense rainbands. It had sustained peak winds of the order of 285 km/hr, and gusts reached as high as 350 km/hr.

Figure 2.5.8.1 Tropical cyclone Monica in the Southern Hemisphere. (Source: NASA web site earthobservatory.nasa.gov/)

Figure 2.5.9.1 Cyclone Asani over the Bay of Bengal and Cyclone Karim over the Indian Ocean viewed by VIIRS on Suomi-NPP on 9 May 2022 (Source: https://www.earthdata.nasa.gov/)

2.5.9 Cyclone Pairs

Full-disc geostationary satellite images quite often show the presence of two cyclonic vortices on either side of the equator at the same longitude. Particularly, when cyclones form over the Bay of Bengal and Arabian Sea, cyclonic storms are seen to develop in the manner of a mirror image on the same longitude in the southern hemisphere.

An interesting situation prevailed on 7 May 2022 when there was a strong westerly wind burst along the equator in the western Indian Ocean. The westerly wind burst, which is associated with the development of a large cluster of deep convective clouds along the equator, led to the formation of tropical cyclone twins, on either side of the equator, one over the Bay of Bengal (Asani) and another over the southern Indian Ocean (Karim). Karim officially formed first in the southern hemisphere followed immediately by Asani in the northern hemisphere. Figure 2.5.9.1 shows at a glance, the two cyclones, Asani rotating anticlockwise and Karim rotating clockwise, on opposite sides of the equator.

Karim and Asani are called twin cyclones not only because they formed at the same time in the same region but also because they were formed primarily from the same parent circulation, the Madden-Julian Oscillation (MJO). On 8 May, Asani further intensified to a severe cyclonic storm, before making landfall as a deep depression system over Andhra Pradesh. It degenerated into a well-marked low-pressure on 12 May. Karim also intensified first but kept moving southwards and eventually dissipated over the cooler ocean.

On a few occasions, cyclones have formed at the same time over the Arabian Sea as well as the Bay of Bengal. However, they are separated by the Indian peninsula and cannot interact directly with each other. Recent examples of such cyclone pairs are of very severe cyclonic storms Gati over the Arabian Sea and Nivar over the Bay of Bengal in November 2020, whose formation was whose influenced by high SSTs and the Madden–Julian Oscillation. Another example occurred in October 2023 when cyclone Tej brewed in the Arabian Sea while cyclone Hamoon emerged in the Bay of Bengal.

There is another type of cyclone pairs that are located on the same side of the equator and not on either side of it. When two such cyclones approach each other, their centres begin to rotate in an anticlockwise manner about a virtual point lying somewhere in between. The two cyclones tend to come closer and spiral into this point, and may even get merged into a single vortex. If one of the vortices is stronger than the other, it will tend to dominate the interaction, and the weaker vortex will orbit around it. This type of interaction is called the Fujiwhara effect, named after Sakuhei Fujiwhara, the Japanese scientist who first postulated it way back in 1921 but in a different context. The Fujiwhara effect is seldom seen in the Atlantic basin but is more common in the Pacific where several tropical cyclones tend to form at the same time.

In April 2021, two tropical cyclones named Seroja and Odette exhibited the Fujiwhara effect over the southeastern Indian Ocean off the coast of Western Australia. Seroja formed first in the Timor Sea on 5 April and Odette formed to its west on 9 April. Odette rotated around the larger Seroja and later merged into Seroja on 11 April. Seroja made landfall on the western coast of Western Australia as a severe tropical cyclone.

2.6 Winter Systems over India

In the months of June to September, India gets its annual monsoon rains and in the post-monsoon months of October and November, tropical cyclones may strike. In comparison, the winter season of December to February is weatherwise a quiet time for India. Occasionally, a tropical storm may develop over the seas, but that is rare. However, at this time of the year, low pressure systems originate over the Mediterranean Sea at frequent intervals and travel across Iran, Afghanistan and Pakistan into extreme northwest India. They are the main weather-producing systems in winter.

2.6.1 Western Disturbances

The winter-time migratory systems enter India from a westerly direction and are therefore referred as western disturbances or western depressions (WDs), depending upon their intensity. Forecasting the approach of WDs is important, as they bring the much-needed winter precipitation over north and northwest India after the southwest monsoon has withdrawn. The stronger WDs also bring in snowfall in the higher mountain regions of Jammu and Kashmir, Himachal Pradesh and Uttaranchal, and may trigger cold waves that penetrate even into central and more southern parts of India. Because of the lack of data and the shallowness of the disturbances, WDs may not always show up clearly in synoptic weather charts except in the case of stronger systems. Satellite cloud imageries are therefore a valuable source of information on the origin and movement of western disturbances.

Bhaskara Rao et al (1971) and Agnihotri et al (1982) studied the approach of WDs into northwest India with the help of satellite cloud pictures. The cloud patterns associated with WDs can be categorized into four broad types:

(a) Systems that develop into extra-tropical depressions have an organized clouding with a banded structure and vortex (Figure 2.6.1.1).

(b) Broad shallow troughs in the westerlies show up as latitudinal bands which extend in a WSW-ENE direction from central Arabia to India and slowly progress eastwards. (Figure 2.6.1.2). They form in association with long waves in the upper tropospheric westerlies or jet streams and some of them may even extend from eastern Sahara to north India.

(c) On rare occasions, meridional cloud bands may be seen, but their movement is erratic and they dissipate on encountering mountain barriers.

(d) All other weaker disturbances are seen as broken amorphous or overcast cloud masses approaching from the west. These cloud masses can be detected in the region 25-35 °N and 40-55 °E from where they move eastwards at an approximate rate of 10° longitude per day.

Figure 2.6.1.3 is a sequence of daily INSAT-3D images from 2 to 5 January 2016 which vividly show the approach of a western disturbance towards India, its gradual movement and dissipation. Srinagar and the Kashmir valley received their first snowfall of the season on account of this system.

Several interesting case studies of individual western disturbances based upon satellite pictures are found in the literature. Sharma et al (1983) have discussed the linkage of a WD with a low in the Arabian Sea lower tropospheric easterlies off the west coast of India, explaining the occurrence of precipitation far to the south.

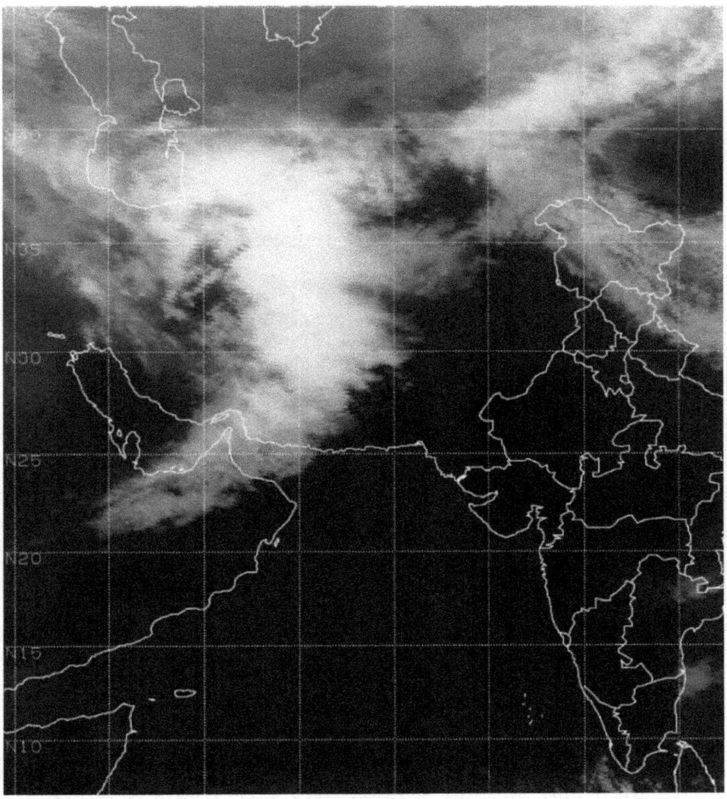

Figure 2.6.1.1 INSAT-3A IR image of 27 January 2004 showing comma-shaped cloud and vortex associated with an intense western disturbance approaching north India. (Source: IMD)

Veeraraghavan et al (1989) studied an active WD that occurred in the month of November and produced widespread precipitation. They also attempted to correlate the satellite-observed cloud top temperatures with the distribution of precipitation produced by the WD.

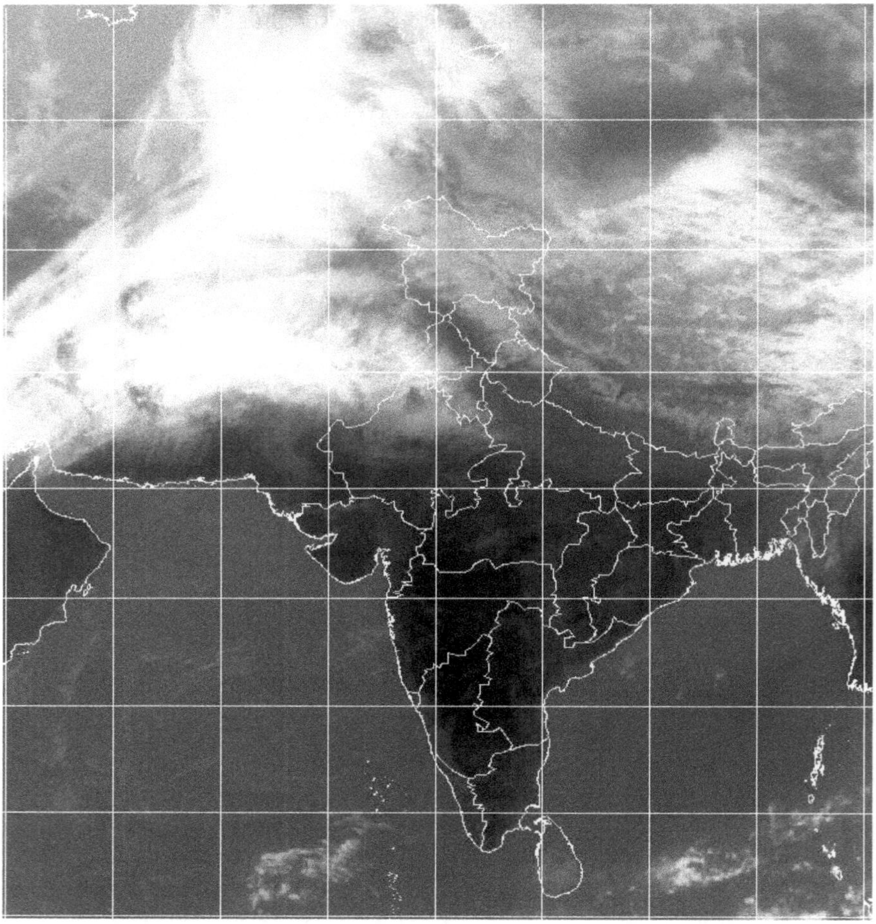

Figure 2.6.1.2 Kalpana-1 IR image of 17 February 2006 showing latitudinal bands associated with a western disturbance over north India and adjoining region. (Source: IMD)

During the winter months, satellite imagery may show extensive formations of high and medium clouds over large parts of north and central India in association with the sub-tropical jet stream (Srinivasan 1971). These clouds are organized into longitudinal bands when the lapse rate is high and wind shear is small, or into transverse bands when the lapse rate is small and there is large wind shear. The clouds are not found on all the days but only when large amplitude troughs are present and moisture advection can occur.

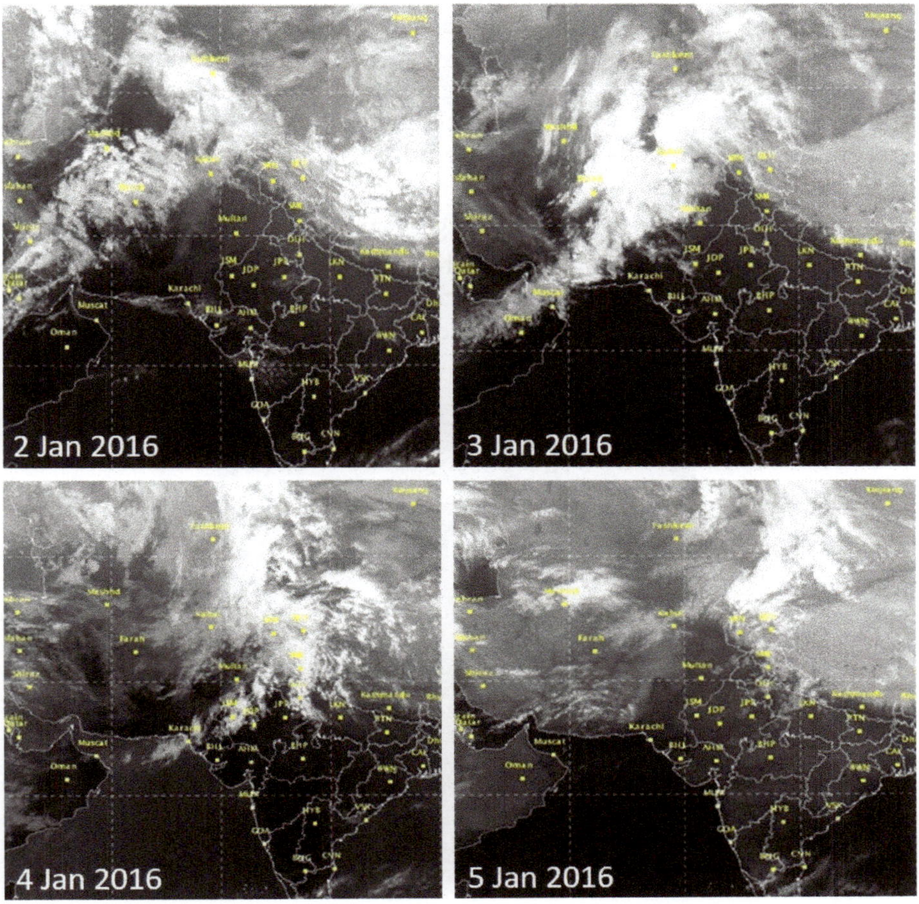

Figure 2.6.1.3 Daily sequence of INSAT-3D TIR1 images for 2-5 January 2016 showing the passage of a western disturbance over north India. (Source: IMD web site imd.gov.in)

2.6.2 Fog and Snow

Winter-time fog is common over northern India, particularly in the wake of a western disturbance that has moved away. Fog is an aviation hazard and it seriously affects aircraft landings and takeoffs, particularly at airports not equipped with instrument landing systems. It also disrupts rail traffic in north India and affects road traffic in major cities. Fog detection, monitoring and prediction is therefore very important.

Widespread winter fog can usually be seen in VIS and SWIR satellite images as an extensive white patch with its northern boundary at the foothills of the Himalayas (Figure 2.6.2.1). It is also possible to do a quantitative detection of fog based on radiances from visible and thermal infrared channels. Nighttime fog detection is usually done with 10.8 and 3.9μ channel brightness temperatures. Suitable thresholds of brightness temperatures have to be chosen and applied in order to identify fog pixels and isolate medium and high clouds and snow. A test for spatial homogeneity is also applied as this is a unique characteristic of fog.

Chaurasia and Gohil (2015) used INSAT-3D data for the winter months of 2013 and 2014 to generate fog maps. They validated their results with surface visibility data available at hourly intervals for five different locations over the Indo-Gangetic plains and found that their technique had a good potential for operational application.

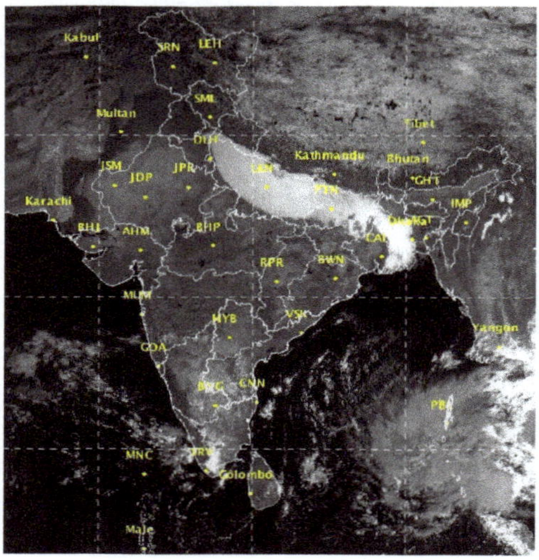

Figure 2.6.2.1 INSAT-3D SWIR image of 6 December 2016 0330 UTC showing extensive morning fog over the plains of northern India south of the Himalayan foothills
(Source: IMD web site imd.gov.in)

In VIS satellite imagery, snow appears bright white because of its very high albedo. In IR channel imagery also snow appears white, but it could be confused with low clouds having temperatures of the same order. Himalayan snow is commonly seen in INSAT images and can be easily identified because of the familiar shape of the Himalaya-Tibet terrain. In case of a cloud cover vitiating the view of the snow, an animation sequence is helpful in differentiating between them. Snow being a surface feature will appear static, while the clouds will move with time. A quantitative assessment of the extent of the snow cover and the snow depth is, however, not a simple matter. It involves a multi-channel analysis of the radiance data over a period of time and the use of statistical relationships (See Section 3.10).

2.7 Extra-tropical Weather

The weather patterns that prevail over the tropical and extra-tropical regions of the globe have significant differences. The main reason is that the quasi-geostrophic approximation which works on the synoptic scale in the extra-tropics does not hold good in the tropics. This means that in the extra-tropical regions, the wind field quickly adjusts itself to the pressure field and can therefore be inferred from it, but this is not so in the tropics where independent analyses of wind and pressure have to be carried out.

Another distinguishing factor is that extra-tropical regions have very strong meridional temperature gradients, resulting in the existence of different kinds of air masses separated by frontal surfaces. The pressure gradients are also steep and there are significant 24 hr changes in the surface pressure. Extra-tropical cyclones keep developing throughout the year as against the preferred seasons of tropical cyclones. The air masses and fronts migrate in an orderly manner and can be tracked on synoptic weather charts and satellite cloud imagery.

2.7.1 Air Masses

An air mass is an extensive body of air that has uniform temperature and moisture properties throughout it in a horizontal plane. Air masses have a source region associated with them, where they acquire their properties. The best source regions for air masses are large flat areas where air can be stagnant long enough to take on the characteristics of the surface below. Large snow-covered polar areas, deserts and tropical oceans, all serve as source regions of air masses. The two most conservative properties of air masses which serve to trace their origin, movement and modification, are the dew point temperature at the surface and the wet bulb potential temperature in the free atmosphere.

Air masses can be classified in several ways. The primary division is with respect to their geographical source region: Arctic (A), Polar (P) and Tropical (T). There is a further sub-division according to the nature of the source region: maritime (m) or continental (c). The resulting combinations are: continental Arctic (cA), continental Polar (cP), continental Tropical (cT), maritime Arctic (mA), maritime Polar (mP) and maritime Tropical (mT). However, cA, cP, mP and mT type of air masses are more predominant than the others.

As air masses move from one region to another they may undergo modifications or acquire additional attributes. Air masses of cA and cP type originate in polar anticyclones, while those of cT and mT originate in subtropical anticyclones. As air masses of the type cA and cP move over oceanic areas, they pick up moisture from the ocean and become maritime air masses mA and mP. If a polar air mass mP moves southward, say from Canada to the southern U.S., it will get warmed over the ground, but it will still remain very dry and get modified into a continental polar air mass cP. The U.S. Gulf Coast region commonly has a tropical air mass in summer. If a dry continental tropical air mass cT moves and stagnates over the mid-western U.S., drought conditions may develop. Maritime tropical air masses mT have their origin in the tropical and subtropical oceans, but move northward transporting heat and moisture. In contrast, continental polar air masses cP originate over the northern plains, and move southward carrying colder and drier air with them.

2.7.2 Cold, Warm, Stationary and Occluded Fronts

As per its traditional definition, a front is a sharp transition zone between two air masses of different density, temperature and humidity conditions. The warm air boundary of the maximum temperature gradient is considered as the actual frontal surface. As fronts extend in horizontal as well as vertical directions, the frontal surface has both horizontal and vertical components. Fronts are named after the type of air mass that is replacing the other. The most commonly observed fronts are the continental Arctic front, maritime Arctic front and the Polar front, separating the cA-mA, mA-mP and mP-mT air masses respectively.

Fronts can be classified into four types:

(a) A cold front is a front that demarcates the leading edge of a cold air mass displacing a warmer air mass

(b) A warm front is the leading edge of a warm air mass displacing a cold air mass.

(c) A stationary front is one in which the air masses are not in motion. If such a front begins to move again, it will once again be called a cold or warm front as the case may be. A noticeable temperature change or shift in wind direction is commonly observed across a stationary front. Cold fronts typically move faster than warm fronts, so they eventually catch up with warm fronts.

(d) An occluded front forms when the two fronts get merged. The cold air undercuts the cooler air mass associated with the warm front, further lifting the already rising warm air.

Fronts can be detected on surface weather charts in many ways. Winds usually converge at the fronts and there are noticeable differences in temperature and pressure on either side of a front. However, fronts extend to higher atmospheric levels as well and they have a vertical structure and slope.

Warm fronts typically have a gentle slope, so the air rises gradually along the frontal surface. This usually favours the development of widespread layered or stratiform cloudiness and precipitation along and to the north of the front. Cold fronts have a steeper slope and air is forced upward more abruptly. This usually leads to a narrow band of showers and thunderstorms along or just ahead of the front, especially if the rising air is unstable.

A dry line is a boundary that separates a moist air mass from a dry air mass. It is also called a dew point front, as sharp changes in dew point temperature can be observed across a dry line. Frontal clouds are a combination of various cloud types such as Cb, Ns, As and Ci. Active frontal zones appear in both VIS and IR satellite imagery as long, multilayered cloud bands extending across thousands of kilometres and having a width of up to a few hundred kilometres depending on their intensity. There are apparent differences in the cloud patterns on both sides of the front.

An active cold front can be easily located in a satellite picture as it lies along the outer segment of the spiral band which forms a cloud vortex. The band narrows down away from the vortex. Cold fronts that are less active, have isolated Cb clouds and appear as narrow, fragmented cloud bands, which look bright in VIS but gray in IR images. Warm front clouds come in different sizes and patterns and a warm front can be clearly identified in satellite images only in the initial stages of vortex development.

In VIS imagery, an occluded front appears as a wide spiral band of bright clouds with a sharp inner boundary. The leading edge of the cloud spiral is more diffuse or ragged. In IR imagery, however, the brightness reduces towards the centre of the cloud spiral where the clouds do not develop to great heights. The cloud band associated with a stationary front is only 200-300 km in width and it does not exhibit any curvature. The cloud mass has an irregular shape and broken boundaries in both VIS and IR imageries.

2.7.3 Mid-latitude Cyclones

A cyclone, whether in the tropics or in the extra-tropics, is an area of low pressure around which the winds flow counterclockwise in the Northern Hemisphere and clockwise in the Southern Hemisphere. However, the characteristics of mid-latitude cyclones are significantly different from those of a tropical cyclone that were described in the preceding sections.

A mid-latitude cyclone in its developing stage is typically accompanied by a warm front pushing northward and a cold front pulling southward, marking the leading edges of air masses coiling around a centre of low pressure, or the centre of the cyclone. The counterclockwise winds around the low pressure area transport warm moist air to its north and cold, dry air to its south. In satellite images, mature mid-latitude cyclones are easily identifiable by the characteristic comma-shaped configuration of the cloud bands associated with them. However, satellite images show four different stages in their life cycle:

(a) *Leaf stage*: During the initial development of a mid-latitude cyclone, the satellite cloud pattern takes the shape of a leaf because of the jet stream pushing into the western edge of the cloud system and spreading the cloud downwind. The highest cloud tops are over the eastern portion of the leaf and middle-level clouds are over the westward portion. The western edge of the leaf is well-defined.

(b) *Open comma stage*: A comma-shaped cloud pattern appears in satellite images when the mid-latitude cyclone has well-developed warm and cold fronts. This is a sign of falling surface pressure. The back edge of the comma cloud represents the position of the cold front where the clouds and thunderstorms form because of the frontal lifting. The front edge of the comma cloud pattern is more diffuse as upper level winds make the clouds to spread out.

(c) *Occluded stage*: As the occlusion forms, the low-pressure circulation separates from the jet stream and the central pressure stops falling. The point at which the occluded, cold and warm fronts come together is referred to as the triple point, and it is often seen in satellite imagery where the jet stream cuts across the system. The cloud pattern is still in the shape of a comma, and the clouds spiral around the centre of the storm located in the comma head (Figure 2.7.3.1).

(d) *Dissipating stage*: As the storm weakens, upper level winds tend to destroy it. The cloud system loses its organized pattern and the comma head gets cut off from the tail, creating cut-off lows that may persist for a long time.

Figure 2.7.3.1 GOES satellite image of 26 October 2010 showing a mature mid-latitude cyclone over the Midwestern U.S.
(Source: NASA web site earthobservatory.nasa.gov)

2.7.4 Sub-tropical Jet Stream

As per its classical definition, a jet stream in the atmosphere is a strong narrow current 1000 km long, 100 km wide and 1 km deep with wind speed greater than 60 knots concentrated along a quasi-horizontal axis. It has wind shears of the order of 10-20 knots/km in the vertical and 10 knots/km in the horizontal direction, featuring one or two vorticity maxima (Berggren et al 1958).

Six types of jet streams are observed commonly in the atmosphere:

(a) the polar night stratospheric westerly jet streams

(b) QBO jet in the lower equatorial stratosphere,

(c) polar front westerly jet stream,

(d) sub-tropical westerly jet stream,

(e) tropical easterly jet stream, and

(f) low level jet stream.

In a non-rotating earth, the warm air rising at the equator will move poleward on both sides. Because of the earth's rotation, the air moving poleward conserves its eastward momentum. The regions around 30° N/S and 50-60° N/S are areas of the highest temperature gradients and also therefore the regions of formation of the polar jet stream and the subtropical westerly jet stream.

The sub-tropical westerly jet stream is generally found at 30 ºN latitude at 200 hPa or 12 km height, but it fluctuates widely in its position and is most intense in winter when the air masses have a stronger surface temperature contrast.

The sub-tropical jet stream is almost a permanent feature of the 300 and 200 hPa level Indian synoptic charts in the winter months of December to February, its mean location being along 27ºN at 12 km height and the mean wind being 100 knots. However, winds as high as 150 and 200 knots may be observed in the jet core and there are instances of the jet stream migrating to a far lower latitude (Joseph 1967).

In IR satellite images, the location of the sub-tropical jet can be identified from the cirrus clouding which predominates on its equatorial side and has a sharp poleward boundary. The cirrus may take various shapes and sizes such as large shields associated with large amplitude troughs, long bands, narrow streaks along the jet axis or transverse bands due to shear.

Satellite images commonly show large masses of cirriform clouds originating in the tropics or the ITCZ and extending poleward. These clouds are usually situated to the south of the wind maximum and represent the meridional transport of momentum and high level moisture from the tropical region (Figure 2.7.4.1).

Figure 2.7.4.1 INSAT-3D TIR image of 2 December 2015 showing long cloud bands associated with jet streams. (Source: IMD web site imd.gov.in)

2.8 Interaction between Tropical and Mid-latitude Systems

The sub-tropical highs insulate the easterly flows in the tropics from the westerly flow of the middle latitudes. However, extra-tropical migratory systems sometimes induce the formation of cold lows in the tropical latitudes to their south. There are occasions when large amplitude troughs develop in the westerlies and extend into the tropical latitudes, which manifest themselves in satellite imageries as long bands surging northeastwards from the tropics. Such an interaction between the tropical and mid-latitude weather systems (abbreviated hereafter as ITML) can occur in many different synoptic settings throughout the year. ITML can lead to an intensification of weather disturbances in winter, outbreak of convection in the pre-monsoon season, oscillations of the monsoon trough and deformation of circulations including cyclonic storms. Figure 2.8.1 is a Kalpana-1 WV image of 23 November 2015 which is a fine example of an ITML episode that resulted in widespread light to heavy rainfall over several parts of the Indian peninsula, particularly Maharashtra and Tamil Nadu.

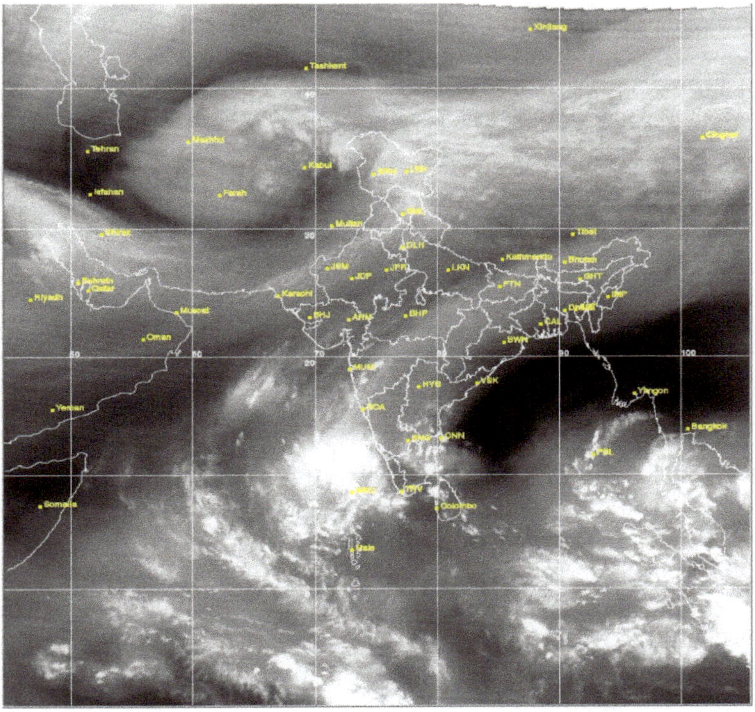

Figure 2.8.1 Kalpana-1 WV Image of 23 November 2015 showing an interaction between easterly and westerly flows (Source: IMD web site imd.gov.in)

Kalsi et al (1992) have shown examples of how satellite imagery provides a means of monitoring ITML in relation to these developments, particularly where the short waves cannot be resolved in conventional weather analysis. In the case of an ITML during a western disturbance, an extension of low level warm and moist easterly flow extends to northwest India and Pakistan, where it meets the mid-latitude cold and dry westerly flow. Though the temperature contrast between the two flows may not be significant, the humidity contrast could be striking. A baroclinic leaf pattern associated with a surface trough of low pressure can be seen in satellite imagery.

In the case of an ITML in the pre-monsoon season, the presence of a westerly trough is noticed by the bands of medium and high clouds forming ahead of it. As it meets the low level southeasterlies over central India, the short waves induce a sub-tropical cyclogenesis over Pakistan which can be seen as comma cloud system. Such an interaction can cause a lot of precipitation over India, mostly in association with thunderstorms.

During the monsoon season, ITML can produce different effects like causing intensification of low level lows or troughs, enhancing rainfall in pre-existing systems, causing recurvature of depressions and lows and even causing a break in the monsoon. It can disturb the orientation of the monsoon trough and alter the regions of heavy precipitation.

Although there have been many cases of tropical cyclones having been sheared off by mid-latitude westerlies, Kalsi et al (1992) have given an example of a marginal cyclone in the Bay of Bengal in October 1988 that got deformed northeastwards as a result of ITML. They have also given a case of abnormally rapid movement of a tropical vortex in the Bay in November 1988 on account of another vortex to its north and its intensification after a merger with it.

2.9 References

Ahmed R., Mohapatra M, et al, 2021a, "An Evaluation of the Advanced Dvorak Technique (9.0) for the tropical cyclones over the North Indian Ocean", *Tropical Cyclone Research and Review*, 10, 201-208.

Ahmed R., Mohapatra M. et al, 2021b, "Characteristic features of Super Cyclone 'AMPHAN'-observed through satellite images", *Tropical Cyclone Research and Review*, 10, 16-31.

Agnihotri C. L. and Singh M. S., 1982, "Satellite study of western disturbances", *Mausam*, 33, 249-254.

Ananthakrishnan R. and Soman M. K., 1988, "The onset of the south-west monsoon over Kerala 1901-1980", *J. Climatology*, 8, 283-296.

Asnani G. C., 2005a, "ITCZ (Inter-Tropical Convergence Zone)", *Tropical Meteorology*, 1, 126-144.

Asnani G. C., 2005b, "Subsidence over equator", *Tropical Meteorology*, Vol. 2, Chapter 5, 42-68.

Asnani G. C., 2005c, "Special features of tropics, monsoons", *Tropical Meteorology*, Vol. 1, Chapter 1, 16-37.

Berggren R., Gibbs W. J. and Newton C. W., 1958, "Observational characteristics of the jet stream – a survey of the literature", *WMO Tech. Note No. 19.*

Bhagat D. K. U. R., 2005, "Balance of kinetic energy in the field of mid-tropospheric cyclone", *Mausam*, 56, 473-500.

Bhan S. K., Paul S. and Kharbanda K. L., 2004, "Cloudbursts in Himachal Pradesh", *Mausam*, 55, 712-713.

Bhaskara Rao N. S. and Moray P. E., 1971, "Cloud systems associated with western disturbances: A preliminary study", *Indian J. Meteor. Geophys.*, 22, 413-420.

Bunker A. F., 1965, "Interaction of the summer monsoon air with the Arabian Sea", *Proc. Symp. Meteorological Results of the IIOE,* WMO, Geneva, 3-16.

Chaurasia, S. and Gohil B. S., 2015,"Detection of Day Time Fog Over India Using INSAT-3D Data", *IEEE J. Selected Topics in Applied Earth Observations and Remote Sensing,* 8, 4524-4530.

Chowdhury A. and Gaikwad S. D., 1983, "On some characteristic features of rainfall associated with monsoon depression over India", *Mausam*, 34, 33-42.

Chowdhury A., Urankar P. S. and Upadhye C. U., 1985, "Cloud characteristics of monsoon depressions as viewed by meteorological satellites", *Mausam*, 36, 491-498.

De U. S., 1970, "Lee waves as evidenced by satellite cloud pictures", *Indian J. Meteor. Geophys.*, 21, 637-642.

De U. S., 1970, "Mountain waves over northeast India and neighbouring regions", *Indian J. Meteor. Geophys.*, 22, 361-364.

De U. S. and Mukhopadhyay R. K., 2002, "Breaks in monsoon and related precursors", *Mausam*, 53, 309-318.

Dvorak V. F., 1975, "Tropical cyclone intensity analysis and forecasting from satellite imagery", *Mon. Wea. Rev.*, 103, 420-430.

Dvorak V. F., 1984, "Tropical cyclone intensity analysis using satellite data", *NOAA Tech. Rep. NESDIS 11,* 47 pp.

Dvorak V. F. and Mogil H. M., 1992, "Tropical cyclone motion forecasting using satellite water vapor imagery", *NOAA Tech. Rep., NESDIS 83.*

Fassulo J. and Webster P. J., 2003, "A hydrological definition of Indian monsoon onset and withdrawal", *J. Climate*, 16, 3200-3211.

Findlater J., 1966, "Cross-equatorial jet streams at low level over Kenya", Meteor. Mag., 95, 353-364.

Findlater J., 1969, "A major low-level air current near the Indian Ocean during the northern summer," *Quart. J. Royal Meteor. Soc.,* 95, 362-380.

Findlater J., 1974, "An extreme wind speed in the low level jet stream of western Indian Ocean", *Meteor. Mag.*, 103, 201-205.

Gadgil S. and Joseph P. V., 2003, "On breaks of the Indian monsoon", *Proc. Indian Acad. Sci. (Earth Planet. Sct.)*, 112, 529-558.

Gokhale M. N., 1971, "Surface wind estimate from satellite pictures of sunglint", *Indian J. Meteor. Geophys.*, 22, 443-446.

Golder R. N., Banerjee S. K. and Debnath G. C., 2001, "Tornado in India and neighbourhood", *Pre-published Sci. Rep. No. 2/2001*, RMC Kolkata, India Met. Dept.

Gupta G. R. and Onkari Prasad, 1991, "Activity of southern Indian Ocean convergence zone as seen in satellite cloud data during pre-monsoon months", *Mausam*, 42, 145-150.

Gupta H. N. and Ghosh S. K., 1980, "North Delhi tornado of 17 March 1978", *Mausam*, 31, 93-100.

Grossman R. L. and Friehe C. A., 1986, "Vertical structure of the southwest monsoon low-level jet over the central and eastern Arabian Sea", *J. Atm. Sci.*, 43, 3266-3272.

Jenamani R. K., 2004, "Distinct synoptic patterns associated with pre-break onset phase and revival of normal monsoon phase", *Mausam*, 55, 591-598.

Joseph P. V., 1967, "A case of very low latitude occurrence of the subtropical jet stream over the India region", *Indian J. Meteor. Geophys.*, 18, 217-226.

Joseph P. V., Raipal D. K. and Deka S. N., 1980, " 'Andhi', the convective dust-storm of northwest India", *Mausam*, 31, 431-442.

Joseph P. V., 1982, "A tentative model of *Andhi*", *Mausam*, 33, 417-422.

Joseph P. V., 1990, "Monsoon variability in relation to equatorial trough activity over Indian and west Pacific Oceans", *Mausam,* 41, 291-296.

Joseph P. V., Fishchild J. K. and Pile R., 1994, "Interannual variability of onset of Indian summer monsoon and its association with atmospheric features, El Nino and sea surface temperature anomalies", *J. Climate*, 7, 81-105.

Joseph P. V. Sooraj K. P. and Rajan C. K., 2003, "Conditions leading to monsoon onset over Kerala and the associated Hadley cell", *Mausam*, 54, 155-164.

Joseph P. V. and Sijikumar S., 2004, "Intraseasonal variability of the low-level jet stream of the Asian summer monsoon", *J. Climate*, 17, 1449-1458.

Joseph P. V., 2006, "SW monsoon convective clouds along the west coast of India associated with the low level jet stream", *National Workshop on Cloud Physics, CESS, Thiruvananthapuram*, 24-25 March 2006.

Joseph P. V., Sooraj K. P. and Rajan C. K., 2006, "The summer monsoon onset process over south Asia and an objective method for the date of monsoon onset over Kerala", *Int. J. Climatol.,* 26.

Joshi P. C. and Simon B., 1994, "NOAA/TOVS derived upper tropospheric temperature changes associated with the onset of southwest monsoon over Kerala coast", *Mausam*, 45, 155-160.

Kalsi S. R. and Halder S. R., 1992, "Satellite observations of interaction between tropics and mid-latitudes", *Mausam*, 43, 59-64.

Kalsi S. R., 1993, "INSAT image analysis of tropical cyclone development in the Bay of Bengal", *Advances in Tropical Meteorology (Ed. Keshavamurty R. N. and Joshi P. C.),* Tata McGraw Hill, 289-300.

Kalsi S. R., 2002, "Use of satellite imagery for tropical cyclone intensity analysis and forecasting", *Met. Monograph, Cyclone Warning Div., No. 1/2002, India Meteor. Dep.*, 83 pp.

Kalsi S. R., 2006, "Orissa supercyclone – a synopsis", *Mausam*, 57, 1-20.

Karmakar S. and Mahbub Alam Md., 2005, "On the sensible heat energy, latent heat energy and potential energy of the troposphere over Dhaka before the occurrence of nor'westers in Bangladesh during the pre-monsoon season", *Mausam*, 56, 671-680.

Kelkar R. R., 1997, "Satellite-based monitoring and prediction of tropical cyclone intensity and movement", *Mausam*, 48, 157-168.

Kelkar R. R., 2012, "Monsoons elsewhere in the world", *Monsoon Monograph.* Government of India, Ministry of Earth Sciences, India Meteorological Department, 1, 96-119

Khole M. and De U. S., 2003, "A study of north-east monsoon rainfall over India", *Mausam*, 54, 419-426.

Khole M. and De U. S., 2004, "Association between the monsoon onset over Kerala (MOK) and sea surface temperature (SST) over north Indian Ocean", *Mausam*, 55, 495-496.

Koteswaram P., 1958, "The Asian summer monsoon and the general circulation over the tropics", Proc. *Symp. Monsoons of the World, New Delhi, 19-21 February 1958*, 105-110.

Krishnamurti T. N. and R. S. Hawkins, 1970, "Mid-tropospheric cyclones of southwest monsoon", *J. Appl. Meteor.*, 9, 442-458.

Krishnamurti T. N., Ardanuy P., Ramanathan Y. and Pasch R., 1984, "On the onset vortex of the summer monsoon", *Mon. Wea. Rev.*, 109, 34-363.

Kulshrestha S. M., 1971, "A satellite study of the Orissa cyclonic storm of October 1967", *Indian J. Meteor. Geophys.*, 22, 313-316.

Litta, A. J., Mohanty U.C. and Bhan S.C., 2020, "Numerical simulation of a tornado over Ludhiana (India) using WRF-NMM model." *Meteorological Applications*: 17, 64-75.

Litta A. J. et al, 2012, "Simulation of tornado over Orissa (India) on March 31, 2009, using WRF–NMM model", *Natural Hazards,* 61, 1219-1242.

Mahajan P. N., 2003, "Satellite data for diagnostics of monsoon disturbances", *Mausam*, 54, 165-172.

Mahajan P. N. and 5 others, 2004, "Proper depiction of monsoon depression through IRS-P4 MSMR", *Proc. Indian Acad. Sci. (Earth Planet. Sci.)*, 113, 223-243.

Manohar G. K. and Kesarkar A. P., 2003, "Climatology of thunderstorm activity over the Indian region: Part I A study of east-west contrast", *Mausam*, 54, 819-829.

Manohar G. K. and Kesarkar A. P., 2004, "Climatology of thunderstorm activity over the Indian region: Part II Spatial distribution", *Mausam*, 55, 31-40.

Manohar G. K. and Kesarkar A. P., 2005, "Climatology of thunderstorm activity over the Indian region: Part III Latitudinal and seasonal variation", *Mausam*, 56, 581-592.

Miller F. R. and Keshavamurty R. N., 1968, "Structure of an Arabian Sea summer monsoon system", *IIOE Meteor. Monograph No.1,* East-West Press, Honolulu.

Mishra D. K., Gupta G. R., Onkari Prasad and Nath T., 1991, "Some aspects of southwest monsoon as seen in satellite cloud imagery", *Mausam*, 42, 261-264.

Mitra A. K. et al, 2018, "Identification of weather events from INSAT-3D RGB scheme using RAPID tool", *Current Sci.* 115, 1358-1366.

Mitra A. K. et al, 2019, "Monitoring of severe weather events using RGB scheme of INSAT-3D satellite." *J. Earth System Sci.,* 128, 36.

Mohapatra M., Sharma M. and Pattanaik D. R.. 2025, "Evolution of cyclone monitoring and forecasting system in India: a review", Mausam,76, 103-124.

Ngara T. and Asnani G. C., 1978, "Five day oscillations in east African low level jet", *Nature*, 272, 708-709.

Olander T. L. and Christopher S. Velden C. S., 2019, "The advanced Dvorak technique (ADT) for estimating tropical cyclone intensity: Update and new capabilities." *Weather and Forecasting* 34. 905-922.

Pai D. S. and Rajeevan Nair M., 2009, "Summer monsoon onset over Kerala: New definition and prediction", *J. Earth System Sci.*, 118, 123-135.

Pant M. C., 1982, Some characteristic features of the low-level jet field over the Arabian Sea during the Indian summer monsoon, *Mausam*, 33, 85-90.

Pearce R. P. and Mohanty U. C., 1984, "Onsets of the Asian summer monsoon 1979-1982", *J. Atmos. Sci.*, 41, 1620-1639.

Prasad K., Kalsi S. R. and Datta R. K., 1990, "Wind and cloud structure of monsoon depressions", *Mausam*, 41, 385-370.

Raghavan S., 1997, "Radar observations of tropical cyclones over the Indian seas", *Mausam*, 48, 169-188.

Rajeevan M., Gadgil S. and Bhate J., 2010. "Active and break spells of the Indian summer monsoon", *J. Earth System Sci.*, 119, 229-247.

Raj Y. E. A., 1996, "Inter- and intra-seasonal variation of thermodynamic parameters of the atmosphere over coastal Tamil Nadu", *Mausam*, 47, 259-268.

Rajeev V. K., Parameswaram K. and Raju C. S., 2006, "Satellite-based Cloud Studies: KALPANA1-VHRR and NOAA-AVHRR", *National Workshop on Cloud Physics,* CESS, Trivandrum, 24-25 March 2006.

Rajeevan M., and Pai D. S. and Das M. R., 2001, "Asymmetric thermodynamic structure of monsoon depression revealed in microwave satellite data", *Current Sci.*, 81, 448-450.

Rajeevan M., Bhate J., Kale J. D. and Lal B., 2006, "High resolution daily gridded rainfall data for the Indian region: Analysis of break and active monsoon spells", *Current Sci.*, 91, 296-306.

Rao R. R., 1990, "Observed variability in the thermal response of the upper north central Arabian Sea to the forcing of onset vortex during summer monsoon experiments", *Mausam*, 41, 439-444.

Rao R. R. and Sivakumar D., 1999, "On the possible mechanism of the evolution of a mini-warm pool during the pre-summer monsoon season and the genesis of onset vortex in the southeastern Arabian Sea", *Quart. J. Royal Meteor. Soc.*, 125, 787-809.

Saha K. R., 1971, "Cloud distribution over equatorial Indian Ocean as revealed by satellites", *Indian J. Meteor. Geophys.*, 22, 389-396.

Sarker R. P. and Chowdhary A., 1988, "A diagnostic structure of monsoon depression", *Mausam*, 39, 9-18.

Sharma R. V. and Subramanian D. V., 1983, "The western disturbance of 22 December 1980: A case study", *Mausam*, 34, 117-120.

Shenoi S. S., Shankar D., and Shetye S. R., 1999, "The sea surface temperature high in the Lakshadweep sea before the onset of the southwest monsoon", *J. Geophys. Res.*, 104, 703-712.

Sikka D. R., 1971a, "Evaluation of the use of satellite photography in determining the location and intensity changes of tropical cyclones in the Arabian Sea and the Bay of Bengal", *Indian J. Meteor. Geophys.*, 22, 305-312.

Sikka D. R., 1971b, "Development of tropical cyclones in the Indian Seas as revealed by satellite radiation and television data", *Indian J. Meteor. Geophys.*, 22, 317-324.

Sikka D. R. and Gadgil S., 1980, "On the maximum cloud zone and the ITCZ over Indian longitudes during the southwest monsoon", *Mon. Wea. Rev.*, 108, 1840-1853.

Sikka D. R., 2006, "Major advances in understanding and prediction of tropical cyclones over north Indian Ocean: a perspective", *Mausam*, 57, 165-196.

Simon B. et al, 2001, "Monsoon onset-2000 monitored using multi-frequency microwave radiometer on-board Oceansat-1", *Current Sci.,* 81, 647-651.

Singh J. and Hem Raj, 1982, "A satellite study of the tropical easterly jet stream during Monsoon-77", *Mausam*, 33, 113-120.

Sinha Ray K. C., 1988, "Satellite observations of mountain waves over northwest India and neighbouring regions and their theoretical verification", *Mausam*, 39, 257-262.

Soman M. K. and Krishna Kumar K., 1993, "Space-time evolution of meteorological features associated with the onset of the Indian summer monsoon", *Mon. Wea. Rev.*, 121, 1177-1194.

Srinivasan V., 1971, "Some case studies of cirriform clouds over India during the winter period", *Indian J. Meteor. Geophys.*, 22, 421-428.

Srinivasan V., Raman S. and Ramakrishnan A. R., 1971, "Monsoon depression as seen in satellite pictures", *Indian J. Meteor. Geophys.*, 22, 337-346.

Suresh R. and Raj Y. E. A., 2001, "Some aspects of Indian northeast monsoon as derived from TOVS data", *Mausam*, 52, 727-732.

Suresh R., Sankaran P. V. and Rengarajan S., 2002, "Atmospheric boundary layer during northeast monsoon over Tamil Nadu and neighbourhood - a study using TOVS data", *Mausam*, 53, 75-86.

Tyagi A., 2007, "Thunderstorm climatology over Indian region", *Mausam,* 58, 189-212.

Tyagi A., Sikka D. R., Goyal S and Bhowmick M., 2012, "A satellite based study of pre-monsoon thunderstorms (Nor'westers) over eastern India and their organization into mesoscale convective complexes", *Mausam*, 63, 29-54.

Tyagi B. et al, 2012, "Thermodynamical structure of atmosphere during pre-monsoon thunderstorm season over Kharagpur as revealed by STORM data", *Pure Appl. Geophys*., 170, 675-687.

Veeraraghavan K. and Nath T., 1989, "A satellite study of an active western disturbance", *Mausam*, 40, 303-306.

Velden C. S., Smith W. L. and Mayfield M., 1984, "Applications of VAS and TOVS to tropical cyclones", *Bull. Amer. Meteor. Soc*., 65, 1059-1067.

Velden C. S., Goodman B. M. and Merrill R. T., 1991, "North Pacific tropical cyclone intensity estimation from NOAA polar-orbiting satellite microwave data", *Mon. Wea. Rev*., 119, 159-168.

Velden C. S., Olander T. L. and Zehr R. M., 1998, "Development of an objective scheme to estimate tropical cyclone intensity from digital geostationary satellite infrared imagery", *Weather and Forecasting,* 13, 172-186.

Waliser D. E. and Gautier C., 1993, "A satellite-derived climatology of the ITCZ", *J. Climate,* 6, 2162-2174.

Wang B., Lin Ho, Zhang Y. and Lu M. M., 2004, "Definition of South China Sea monsoon onset and commencement of the east Asia summer monsoon", *J. Climate*, 17, 699-710.

Webster P. J. et al, 1998, "Monsoons: processes, predictability and the prospects for prediction", *J. Geophys. Res*., 103, C7, 14451-14510.

Chapter 3

Estimation of Atmospheric, Oceanic and Land Parameters

In its early years, satellite meteorology remained primarily confined to the observation of clouds and cloud patterns in satellite images. The information so derived was used as a value addition to the synoptic weather plotted by operational meteorologists the world over. Cloud imagery became a popular and extensively used product that could reveal a lot of information at a glance. It was soon realized, however, that the satellite-measured radiances could be used to estimate several atmospheric, oceanic and land parameters from either physical considerations or by developing suitable statistical techniques. Satellites which were intended essentially to serve non-meteorological applications also began to be considered for extracting useful meteorological information. Parameter retrieval became a new and fast growing branch of satellite meteorology because of two advantages that satellites offered over *in situ* measurements. One was that they covered the ocean regions and the other was that fresh retrievals can be made with every satellite pass.

3.1 Cloud Motion Winds

Historically speaking, the wind vane and the anemometer were two of the earliest meteorological instruments to be developed. Wind cannot be seen, but it can be made to move objects and its energy can be harnessed. In 1806, Admiral Sir Francis Beaufort developed a rating system for an accurate visual estimation of wind speed just by closely observing the movement of objects that we may commonly see around us. This system was originally developed for use by sailors but it was later modified and extended for use on land. It became well-known as the Beaufort Scale, and what is important about it is that it has stood the test of time. In fact, the 64 knots wind speed that occupied the highest place on the 12-point Beaufort Scale is regarded as the hurricane speed threshold even in today's satellite age.

Early in the 20th century, pilot balloons came into widespread use for measurement of upper air winds. They are simple and easy to deploy. A small balloon is filled with gas and released and as it ascends, its azimuth and elevation are measured with an optical theodolite by an observer on ground. The track of the pilot balloon is plotted and the winds at various standard levels are computed from its horizontal displacement. Since the balloon moves very fast initially, it is difficult to make measurements in the boundary layer which are required for many applications. If the balloon gets into a cloud, it is lost from view, which is a serious constraint during the monsoon season because the sky is generally cloudy. At very high levels, the balloon may burst or the observer may lose its track. Night-time ascents are possible by attaching Chinese lanterns to the balloons, but tracking is not easy.

The rawinsonde is similar in principle to the pilot balloon but it employs a larger balloon with an attached reflector which is tracked with a radio-theodolite. As the balloon does not burst easily, winds can be obtained up to a height of 30 km or even higher. In the radiosonde, an electronic package with a transmitter is attached to the balloon. The sensors measure the pressure, temperature and humidity at various levels as the balloon rises. The signals are received at a ground station and decoded to retrieve the vertical profiles of temperature, humidity and other parameters.

Most upper air stations take two to four pilot balloon ascents and two radiosonde ascents per day at 0000 and 1200 UTC. The observations are expensive as manpower and consumables are involved. The radiosonde packages are usually lost and the few that are retrieved are generally not in a reusable condition. The global upper air observation network is therefore not easy to maintain. Further, it is essentially a land network as radiosonde ascents over the sea are made by very few ships, and upper air data is absent over most parts of the oceans.

Beaufort's method of estimating the speed of wind at the surface from the movability of objects around us, is no longer in vogue as we have anemometers and wind vanes to do that job. However, his basic principle is still used for deriving upper air winds by tracking the movement of balloons. After satellites started sending images of clouds over the earth, the application of Beaufort's principle got further extended to the extraction of winds at various levels by calculating the displacement of clouds and their direction of movement between two successive satellite images.

3.1.1 Extraction of CMWs

The ATS series of geostationary satellites, launched in the late sixties, provided for the first time, an opportunity to determine wind velocities by tracking cloud movements over a time series of satellite pictures (Fujita 1969). Since then, techniques for derivation of Cloud Motion Winds (CMW) have undergone continuous refinement and CMWs are now being derived operationally over extensive areas within the coverage of U.S., European, Indian and Japanese geostationary meteorological satellites. Menzel (2001), Le Marshall et al (2003), Khanna et al (2000) and Bhatia et al (2002) have described the evolution of the CMW derivation techniques over the years.

These products have become known by various names such as cloud drift winds, cloud motion winds, cloud motion vectors, and atmospheric motion vectors.

The derivation of CMWs with satellite imagery can be done best with geostationary satellite imagery. Generally, a set of three consecutive half-hourly images is used to track the motion of clouds, either visually by an expert or by subjecting the images to an automated procedure on an image processing system.

In the manual method, which was used in the early years, suitable clouds or cloud features, called targets or tracers, were visually identified and their movement was monitored across successive images. In the automated method, an array of pixels is selected in the first image instead of a single cloud or feature. The target is defined in terms of characteristics such as brightness temperature gradients and spatial coherence. An auto-correlation technique is employed to search subsequent images to identify cloud patterns similar to the target (Kelkar et al 1986, 1993).

In both the methods, the cloud displacement is measured and divided by the time interval between images to get the wind speed, and the direction is obtained from the geometric orientation of the displacement vector. The derived CMW has to be assigned to a proper height and this is usually done with reference to the cloud top temperature, which is obtained from the IR images.

The manual method was very time-consuming but an analyst could derive very accurate vectors and eliminate spurious ones because of personal experience and knowledge of the synoptic situation. On the other hand, automatic pattern-matching techniques can yield thousands of vectors in a much shorter time, but they have to be subjected to stringent quality control in which a large number gets eliminated.

Automated wind extraction methods (Holmlund 1998) involve the computation of statistical correlations between target and search patterns. These correlations can never be very high because of various factors such as navigation and registration errors, brightness changes because of change in solar illumination or vertical growth of cloud, or changes in the cloud form and size. Many cloud features may get altered between successive images.

Image enhancement techniques can be employed to mask noisy signals and isolate a particular level or area. Filter functions can be used to detect distinctive signals and the pattern peaks can then be tracked. The quality control process involves checking for consistency over successive pairs of images, continuity with the previous retrievals and conformity with nearby vectors within the set. Additional checks are carried out by cross-checking with *in situ* wind observations if available, comparing with numerical model forecasts, or by reference to climatology. Quality control is a vital part of the operational CMW extraction process (Holmlund et al 2001).

All types of imagery can be used for CMW derivation but they have their relative advantages and disadvantages. Visible images generally have higher resolution compared to other channels and the cloud features are more clearly defined. The low clouds and thunderstorms are predominant but their brightness changes with time. Winds can be extracted from VIS images only during daytime. This limitation does not apply to IR images, but their resolution is not as good and the cloud features are less sharp. High clouds are predominant in IR images and they mask the middle and lower clouds. The cloud top temperature may change during the derivation if the cloud grows vertically. In WV images, large-scale moisture patterns can be tracked, but these are mainly in the middle troposphere.

One of the difficult parts of the CMW extraction process is the assignment of a proper height to the vector although various methods have been tried for this purpose (Nieman et al 1993). An appropriate height is usually found with reference to the cloud top temperature and applying a vertical temperature profile. Such a profile may not be available, particularly over the ocean. Other difficulties arise from the fact that the target pattern may consist of a mixture of clouds at different levels or the target cloud could also be gaining height or dissipating. In case of low level cumulus clouds seen more in VIS images, the CMW usually corresponds to the cloud base at 900 or 850 hPa. Middle level clouds are still more difficult to handle.

The assumption that a cloud is drifting with the prevalent wind may itself be invalid under certain conditions, for example over mountains. Therefore, all cloud movements cannot be assumed to have been driven by the wind flow.

The basic requirement of CMW retrievals is of course the presence of clouds and their suitability to serve as tracers. However, over areas of clear skies, CMWs can be retrieved by tracking the movement of moisture patterns in WV imagery, usually above the 400 hPa level. The same height assignment method that is used in IR can be applied to WV tracking also, but not if the areas are totally cloudfree.

Generally, VIS imagery produces better low level winds, and IR or cloudy WV imagery gives better upper level winds. VIS imagery gives a clearer distinction between the ocean surface and the low level clouds than does IR imagery. It is obviously not possible to derive CMWs right on the sea surface. This can be obtained only from scatterometers (see Section 4.3).

After the launch of INSAT-3D which has a 6-channel imager, extraction of water vapour winds has become possible. New algorithms have also been introduced for the extraction of CMWs from the data of the two thermal channels.

Attempts have been made to overcome the limitations in the derivation of night-time CMWs when VIS images are not available, by using the GOES SWIR 3.9 μ channel as a new alternative (Velden et al 2005). This is a slightly cleaner window channel than the TIR, making it more sensitive to warmer or lower tropospheric temperatures and it is less sensitive to cirrus clouds. Both these factors make it possible to extract as much as 40 % more low level CMWs at night.

A limitation of geostationary satellites is that they do not see the polar regions and cannot therefore be used to extract winds poleward of the 65° latitudes. Velden et al (2005) have described an attempt for deriving CMWs at such high latitudes from polar orbiting satellite imagery. Polar orbiters have orbital periods of about 100 min, but they can see the polar regions in every orbit, providing overlapping images, from which CMWs can be extracted by the same technique as that used for geostationary satellites images. MODIS IR 11 μ and WV 6.7 μ band images have been successfully used for tracking clouds and cloud features in the polar regions (Key et al 2003). After the launch of the two MetOp satellites in 2011-12, Atmospheric Motion Vectors (AMV) are being derived operationally and a new global AVHRR winds product has now become available (Borde et al 2016).

3.1.2 Use of Rapid-Scan Imagery

Successful tracking of features such as cumulus clouds over land, which have lifetimes that can be considerably less than 30 min, demands the use of imagery with time intervals of the order of 5-15 min. In the U. S., rapid-scan GOES satellite imagery has been used in operational forecasting since the 1980s. Forecasters can utilize the additional detail that can be captured from more frequent imaging in events associated with rapidly changing cloud structures. A study by Velden et al (2000) showed that the more frequent image sampling improved the VIS and IR vector field density and quality, and the optimal tracking time interval was 5 min for VIS, 10 min for IR and 30 min for WV winds.

GOES imagery can be updated every 15 min over the continental U. S. or every 7.5 min over a smaller area, and even every minute for monitoring severe weather and tropical cyclones. New strategies have been developed for operational CMW derivation to take advantage of such rapid- and super rapid-scans from GOES (Velden et al 2005). Use of the 1 min super rapid-scan requires sophisticated image processing and tracking in order to remove navigation and registration errors (Hasler et al 1998).

3.1.3 Applications of CMWs

In recent years, the global radiosonde network has been under pressure and subjected to severe reductions, primarily due to budget constraints and operational challenges. Radiosonde stations in many countries now take only a single daily ascent and some island stations have been closed down altogether. As a result, satellite derived cloud motion winds have become indispensable for initialization of numerical models and for many practical applications such as aircraft flight planning, CMWs provide upper wind data at regular time intervals globally, whereas conventional radiosonde observations are sparse in both space and time, particularly over the oceans. A lot of research and development work goes on at numerical weather prediction centres towards the assimilation of new satellite wind products and improving the way the CMW data is used.

Figure 3.1.3.1 Low level cloud motion winds over Arabian Sea and north Indian Ocean on 16 August 2025 at 09 UTC derived from Meteosat-9 data. (Source: CIMMS web page https://tropic.ssec.wisc.edu/tropic.php)

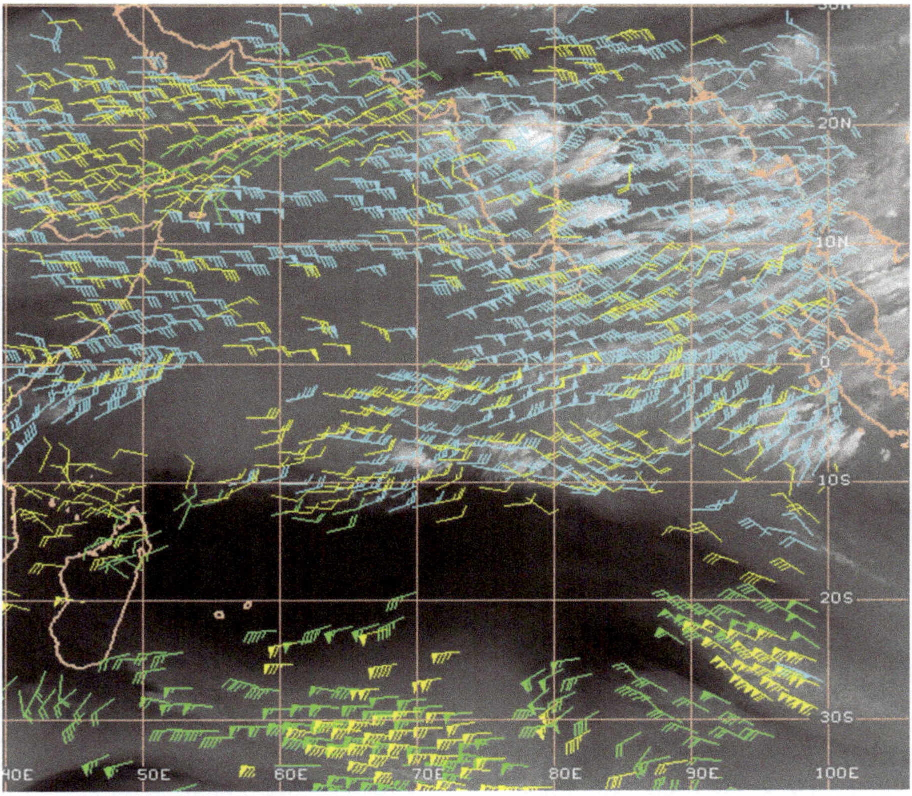

Figure 3.1.3.2 Upper level cloud motion winds over Indian subcontinent and adjoining seas on 16 August 2025 at 03 UTC derived from Meteosat-9 data. (Source: CIMMS web page https://tropic.ssec.wisc.edu/tropic.php)

CMWs over the north Indian Ocean have been found to be particularly valuable for understanding the low level air flow during the Indian summer monsoon season. They are able to capture vital information on the formation of eddies, cross-equatorial flow, approach of the southwest monsoon towards the west coast of India, off-shore vortices and the activity of the two branches of the monsoon over the Arabian Sea and the Bay of Bengal (Yadav et al 1989), which cannot be obtained from conventional data. Sant Prasad et al (1990) compiled average monthly mean patterns of INSAT CMWs that bring out the evolution of the wind flow from the pre-monsoon to the monsoon months. The cross-equatorial flow and the low level jet, and the easterlies over north India and the upper level tropical easterly jet, all typical of the summer monsoon season, are clearly seen in the CMW fields in Figures 3.1.3.1 and 3.1.3.2 respectively.

Mitra et al (2015) have made a quantitative analysis of water vapour winds derived from Kalpana-1 and their role in the initialization of numerical models while Deb et al (2010) have studied their impact on Indian Ocean tropical cyclone forecasts.

3.2 Outgoing Longwave Radiation

The Outgoing Longwave Radiation (OLR) at the top of the atmosphere is an important component of the radiation budget of the earth-atmosphere system. However, in recent years, satellite-derived OLR has also been found useful as a proxy for other meteorological parameters which cannot otherwise be observed globally. In the process of derivation of OLR the minor variations in cloudiness get averaged out, and the OLR values can be plotted as a field and subjected to a spatial and temporal analysis which is not possible with simple cloud imagery. Most important is the strong relationship that has been found to exist between OLR and convective activity and large-scale precipitation over the globe. Low OLR usually indicates cold surface temperatures or presence of clouds, while high OLR indicates warm surface temperatures and absence of convection. The Inter-Tropical Convergence Zone (ITCZ), a region of persistent thunderstorms in the vicinity of the equator is marked by low values of OLR. Major deserts and oceanic subtropical highs have the largest OLR during summer.

3.2.1 ERBE and GERB Measurements

Since 1978, various satellites have carried instrument payloads for measurement of both solar and longwave radiation, Nimbus-7 being the first satellite to carry an Earth Radiation Budget (ERB) payload. A composite Earth Radiation Budget Experiment (ERBE) was designed around three orbiting satellites, ERBS, NOAA-9 and NOAA-10 over the period 1984-1990. The first ERBE instrument aboard the NASA Earth Radiation Budget Satellite (ERBS) was launched from the Space Shuttle Challenger in October 1984. Two more ERBE instruments were launched on NOAA-9 and NOAA-10 satellites in 1984 and 1986 respectively. The main objective of the ERBE mission was to provide accurate measurements of incoming solar energy and shortwave and longwave radiation reflected or emitted from the earth back to space. The other goals of ERBE were to understand the radiation balance between the sun, earth, atmosphere and space which drives our weather and climate system and to establish an accurate, long-term baseline dataset for studying climate system changes (Ramanathan et al 1989, Wielicki et al 2002).

The ERBE Scanner had three scanning detectors of the thermistor bolometer type, in the shortwave 0.2-5 μ, longwave 0.5-50 μ and total 0.2-50 μ wavelength regions. ERBE also included a non-scanning instrument which had five detectors. Of these, four operated in a nadir staring mode and one was a solar monitor that was used only for solar calibration measurements. Two of the detectors had a wide field of view and two had a medium field of view, one each for the total and shortwave bands. The Solar Monitor detector was an active cavity pyrheliometer sensitive to all incident irradiance.

Until 2002, earth radiation budget observations were made from instruments flown in low earth orbit, which provided good spatial resolution because of the proximity to the surface, but had a relatively coarse temporal sampling. This limited their usefulness for studying short-lived events and features such as convective clouds, frontal systems, and release of aerosols associated with dust storms or volcanoes. The aim of the Geostationary Earth Radiation Budget (GERB) project was to provide the accurate, rapid measurements that are required to study the forcing and feedback mechanisms on the shorter time scales that are important in many cases, and in turn, to use improved understanding of these mechanisms to determine inter-annual and longer-term climate variability. The Meteosat Second Generation (MSG) satellites carried a GERB instrument on board.

The GERB instrument (Harries et al 1999, 2005) consisted of an optics unit and an electronics unit. The optics unit contained the imaging optics and detector system, a de-spin mirror, and a quartz filter, along with two onboard calibration targets comprising the thermal blackbody source and the shortwave calibration monitor. Alternate scans measured either the TOTAL spectrum 0.32-100 μ of radiation from the earth, or only SW wavelengths shorter than 4.0 μ. The LW measurement was obtained as a difference between coincident TOTAL and SW.

The GERB instruments have been continued with subsequent Meteosat series satellites (see Section 7.1.1).

3.2.2 OLR Estimation

It is possible to estimate OLR from narrow-band thermal infra-red (TIR) window measurements made by polar orbiting as well as geostationary meteorological satellites. Although OLR is by definition a broad band parameter (4-80 μ), and meteorological satellites measure the radiance in a very narrow window (10.5-12.5 μ), it is possible to fit a statistical regression between the satellite radiance and OLR. While physically this amounts to inferring the whole from a part, in practical terms the method has been found to yield useful information (Ohring et al 1984).

In terms of temperature and assuming a nadir view, the relationship between OLR and radiance relationship is of the form

$$T_f = T_r \, (a + b \, T_r)$$

where T_r is the brightness temperature in °K measured by the satellite, T_f is the computed flux temperature in °K, and quantities a and b are constants. OLR is then computed from T_f as σT_f^4, where σ is the Stefan-Boltzmann constant (see Section 1.1.3).

The regression equation for the INSAT-1B VHRR which had an IR channel 10.5-12.5 µ was derived using the spectral response characteristics of this window channel. The constants a and b have the values 1.1889 and 0.000989 /°K respectively for zero zenith angle. For increasing zenith angle, both a and b decrease in magnitude (Rao et al 1989). For obtaining the regression equation, radiative transfer computations were performed on an array of atmospheric profiles to simulate the radiances measured by the satellite radiometer within its wavelength range at different viewing angles. The process was repeated for the entire longwave region to obtain the total flux at the top of the atmosphere. The two sets of flux data were then subjected to regression analysis which yielded the values of the constants a and b.

The procedure for the computation of OLR assumes a knowledge of the brightness temperature of each individual pixel in an INSAT image. The mean OLR can then be obtained in two ways, either by calculating the OLR for each pixel as per its brightness temperature and then averaging the OLR values over a 2.5 × 2.5 ° latitude/longitude box, or by calculating the mean brightness temperature for the box and getting the corresponding OLR. However, the two approaches return very similar values of the OLR.

Arkin et al (1989) found that NOAA OLR values systematically exceeded those derived from INSAT-1B by 5-10 w m^{-2} over land, but were lower than or equal to the INSAT-1B values over cloudy ocean regions. One of the reasons is that the INSAT-1B OLR averages were calculated from radiances sampled every three hours while NOAA OLR averages were based on only two satellite passes with equator crossing times of 0230 and 1430 local time. Kelkar et al (1993) have examined the diurnal cycle of OLR by computing the average INSAT-1B OLR values separately at eight different hours of the day. Thus while the OLR derivation methodology is the same whether the radiometer is on a polar or geostationary satellite, the sampling frequency of the radiometer is much higher in a geostationary satellite and its OLR average will be more representative and realistic.

For estimating OLR from the INSAT-3D imager which has two TIR and one WV channel, a new approach has been adopted (Garg 2014) and the equation is of the form

$$\mathrm{OLR} = F\,[\theta,\ N_{i,}(\theta)]$$

where θ is the satellite zenith angle, N_i is the observed narrowband radiance and i is channel number. The regression coefficients have been determined by performing a radiative transfer model simulation under various atmospheric conditions. For this purpose a large database of spectral radiance fields was built using the Santa Barbara DISORT Atmospheric Radiative Transfer (SBDART) model. The radiative transfer computations were performed at 19400 wavelengths covering the spectral region 3-100 μ for 4700 different vertical profiles. A genetic algorithm was used to develop an optimum regression equation for estimating the OLR.

Yadav et al (2022) analyzed the INSAT-3D OLR data over 7 years (2014–2020) and compared it with CERES (see section 7.1.1) data and found them to be in reasonably good agreement. They also carried out a case study of OLR values over Cyclones Amphan and Vayu.

3.2.3 Applications of OLR

The satellite-derived OLR is a composite parameter that is modulated by four factors: (a) land or sea surface temperature, (b) cloudiness, (c) vertical temperature structure of the atmosphere and (d) optical depth of atmospheric water vapour. It can therefore be used as a proxy for convection and large-scale precipitation and has found application in many areas. Global OLR data sets are now available over periods long enough for means and anomalies to be computed.

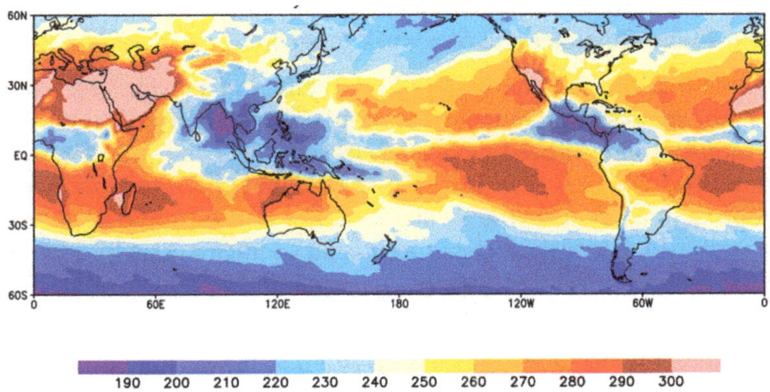

Figure 3.2.3.1 Global monthly average OLR in w m^{-2} for June 2025 derived from NOAA-18 AVHRR IR channel radiances.
(Source: https://www.cpc.ncep.noaa.gov/)

Figure 3.2.3.1 shows the global distribution of monthly average OLR for the month of June 2025 in which the Indian monsoon was advancing aggressively over the subcontinent. The isopleth of 240 w m^{-2} generally delineates the region of monsoon rains and the lowest OLR values (blue areas) are found over the Bay of Bengal and the Indian peninsula region into which the monsoon had advanced.

An important operational use of INSAT OLR is in the declaration of the arrival of the southwest monsoon over Kerala. The deciding criterion is that the OLR should be below 200 w m^{-2} over the region bound by latitudes 5 to 10 °N and longitudes 70 to 75 °E.

Pattanaik et al (2005a) studied the contrasting features associated with an early and late onset of the monsoon over Kerala. In 2004, the southwest monsoon arrived over Kerala on 18 May, 14 days ahead of normal, while in 2003, it had arrived on 8 June, a week late. Using NOAA OLR anomalies, they found that in the early onset year of 2004, the pre-monsoon convective activity was stronger over the Indian region than over the South China Sea and western Pacific, while in the late onset year of 2003, the situation was quite opposite. The development of a strong cross-equatorial flow off the Somali coast in early May was associated with early onset, whereas the cross-equatorial flow remained throughout May and early part of June in the year of late onset.

Srivastava et al (2002 and 2004) who analyzed NOAA OLR data over the period 1979-2000 found a significant negative correlation between the OLR anomaly over north Atlantic Ocean in the month of May and the monsoon rainfall over northwest India in June-September. When May OLR anomalies are positive over the north Atlantic Ocean, June-September sea level pressure anomalies are positive and anomalous 850 hPa easterlies prevail over peninsular India, characteristic of a weak monsoon.

In a similar study, Pattanaik et al (2005b) attempted to derive a relationship between the Indian summer monsoon rainfall and convective anomalies during the pre-monsoon and preceding winter seasons over the Indo-Pacific region based upon NOAA OLR data for 1975-2004. They observed that winter-time convection anomalies over the western Pacific slowly move northwestwards and get established over the Indian Ocean and southeast Asia by May in a year of excess monsoon rainfall, while in years of deficient rainfall, such a migration does not occur.

In an attempt to bring in more objectivity into the definition of the dates of withdrawal of the southwest monsoon, Shyamala et al (1999) used INSAT-derived OLR values, daily as well as pentad, for the month of September for the years 1987-1993 over the area covering northwest India and north Arabian Sea. They found that the OLR increases steadily towards the end of the monsoon season and when it crosses the threshold of 260 w m^{-2}, and persists at that value for 3 days, it is a signal of monsoon withdrawal. However, they have cautioned that this objective criterion should be applied only within a week of the normal date of withdrawal to eliminate spurious signals due to temporary breaks or weak monsoon conditions. Shyamala (1997) found OLR data to be useful in a similar manner for tracking the onset of the southwest monsoon and evolving objective criteria for determining the onset. The OLR values over the southeast Arabian Sea and Kerala coast (5–10°N and 70–75°E). have to be below the 200 wm^{-2} threshold to confirm the presence of deep convection associated with the monsoon's onset. Krishnan et al (2000) have classified break monsoon days in terms of OLR anomalies over the Bay of Bengal and central and northwest India.

Gadgil et al (2003) found that during rain breaks, the OLR anomalies from the July-August OLR mean over the Indian region are positive over most of the Indian region, exceeding 30 $w\ m^{-2}$ over a large area. They also identified a unique feature associated with the breaks in the monsoon whether they are defined in terms of rainfall or circulation. This is a so-called quadrupole in the OLR anomaly pattern over the Asia-west Pacific region, arising from OLR anomalies having opposite (same) sign as anomalies over the Indian region occurring over the equatorial Indian Ocean and northern tropical (equatorial) parts of the west Pacific. This suggests that convection over the Indian region is linked to that over the east Pacific not only on the inter-annual scale but on the intra-seasonal scale as well.

Joseph et al (2004) have studied the intra-seasonal variability of the low level jet stream (LLJ) over the Arabian Sea associated the southwest monsoon using NCEP 850 hPa wind reanalysis and NOAA OLR data for 1979-1990. They constructed a OLR-wind composite for the pentad around the date of onset of the monsoon over Kerala. Similar composites were also prepared for active and break spells of monsoon activity. They concluded that in active monsoon conditions the core of the LLJ runs across peninsular India at 15 °N but during a break it moves southward to 10 °N.

3.3 Large-Scale Precipitation

In the early years of satellite meteorology, there were many attempts to derive some kind of rainfall information from the clouds seen in satellite imagery, particularly from the point of view of forecasting flash floods. The basic problem here is that knowing only the reflectivity or temperature of the top of a cloud, is not sufficient to infer what kind of vertical structure it would be having below it, and it is all the more difficult to assign a rain rate to it. Then again, in images with large pixel sizes, the estimated rainfall cannot be compared to point measurements made by rain gauges. Many of the techniques that were developed initially were quite empirical and had location-specific applications. Some techniques were more useful for climatological purposes. A critical survey of the early satellite-based rain estimation algorithms has been made by Adler et al (1988).

3.3.1 GOES Precipitation Index

The GOES Precipitation Index (GPI) is derived from a simple technique which uses only the cloud top temperature measured by geostationary satellites like GOES (Arkin and Meisner, 1987). It has also been applied to INSAT (Arkin et al 1989, Rao et al 1989) and other geostationary satellites. The technique uses IR images taken every three hours and assigns the pixels to 2.5° × 2.5° lat/long boxes. Several hundred pixels fall in a given 2.5° × 2.5° box and the temperatures of all the pixels in a box are distributed over a 16-class histogram covering temperature ranges of less than 190 °K, 191-200, 201-210, 211-215, 216-220, ….., 261-265, 266-270 and higher than 270 °K. The mean temperature of the box and variance are computed. In the original method, the histograms were accumulated over a month, but the accumulation can as well be done over a day, pentad or week if the number of pixels is sufficient for statistical analysis.

The GPI rainfall estimate over the period is derived as

$$R = F \times C \times T$$

where R is the areal rainfall estimate in mm,

F is the fractional coverage of IR pixels colder than a certain threshold (typically 235 °K) within each 2.5° × 2.5° box,

C is an assumed constant rain rate of 3 mm/hr or 72 mm/day and

T is the number of hours over which F was compiled.

The GPI technique was originally developed for the north Atlantic region during the GATE experiment of 1979 and the value of the constant C was derived empirically as the result of a linear regression between GOES IR data and co-located radar observations, which is essentially location-specific. The threshold temperature of 235 °K was also chosen empirically. Besides these inputs, the technique does not incorporate any physical processes related to precipitation or clouds.

The GPI technique has been widely used for rainfall estimation in the tropics and parts of the extra-tropics where the surface temperature is relatively warm and the rainfall is mostly convective in nature. The estimates are not always accurate over land and particularly the mountain regions, where rainfall is more orographic than convective. The value of the constant 71.2 mm/day may also need to be recalibrated for land regions using radars and dense rain gauge networks that may be available. The method does not work well in regions of persistent non-precipitating cirrus clouds colder than the threshold of 235 °K. The threshold may also not hold good where the rainfall is originating predominantly from warm clouds. Finally, it must be remembered that the method is meant only for large-scale averaged rainfall and cannot be adapted for purposes of short-period heavy rainfall or mesoscale analysis where the flattening out of rainfall maxima is not desirable.

In spite of all such weaknesses in the GPI technique, the great advantage of the technique is in its simplicity and in the use of only IR data which is available frequently over most areas of the globe from geostationary and polar orbiting satellites. Moreover, many studies (Janowiak 1992, Kelkar et al 1990) have shown that the GPI technique does in fact yield useful information about rainfall over the oceans and other land regions where radar and rain gauge data are sparse or not available.

3.3.2 Rainfall Estimation from Microwave Sensors

Passive microwave channel measurements from satellites provide another means for estimation of large-scale rainfall. Microwaves, on account of their longer wavelengths, are able to penetrate clouds and interact strongly with the raindrops present in clouds. Therefore, while techniques like GPI that use TIR channel data are primarily statistical, the process for rainfall estimation from microwave data has a physical basis.

Nimbus-5 and Nimbus-7 were the earliest satellites to carry passive scanning microwave radiometers and clearly demonstrated the potential of microwave measurements. The Special Sensor Microwave Imager (SSM/I), was first flown on the DMSP satellite in 1987. The TRMM Microwave Imager (TMI) on board the TRMM satellite, was another passive microwave instrument operating at 5 microwave frequencies.

As the microwave emissivity of water is about 0.5, ocean surfaces look very cold to a passive microwave radiometer. Raindrops, on the contrary, appear to have a temperature that is almost equal to their physical temperature and so can be easily distinguished against the cold water surfaces. Land emissivity on the other hand is in the range of 0.9 to 1 and it becomes difficult to differentiate between land and rain as both have almost the same brightness temperature. Moreover, land surface emissivity has a geographical variation because of the changing vegetation, presence of snow cover and variations in topography. Certain properties of rainfall, however, can still be inferred even over land.

There are different approaches to microwave rain estimation which make use of three particular properties of microwave radiation. First, compared to the background atmosphere and ocean, the emission of rain clouds is much higher and it increases with the rain rate. Second, high frequency microwaves (85.5 GHz) are strongly scattered by ice nuclei present in many raining clouds. This reduces the microwave signal at the satellite and offers a contrast against the warm land background. Third, the polarization diversity in microwave measurements, arising due to polarization from the surface and de-polarization from hydrometeors, can be used to advantage.

The largest single source of error in microwave measurements is beam filling. Other limitations are introduced by the low sampling frequency of orbiting satellites and the discontinuous nature of precipitation. It is further difficult to place passive microwave sensors at geostationary height. Even then, because of their all-weather capability, passive microwave data are being used not only for precipitation estimation but in a variety of applications to retrieve geophysical parameters like ocean surface wind speed, area covered by ice and characteristics of ice, snow cover, cloud liquid water, integrated water vapour, soil moisture, land and sea surface temperature (Gairola et al 2003).

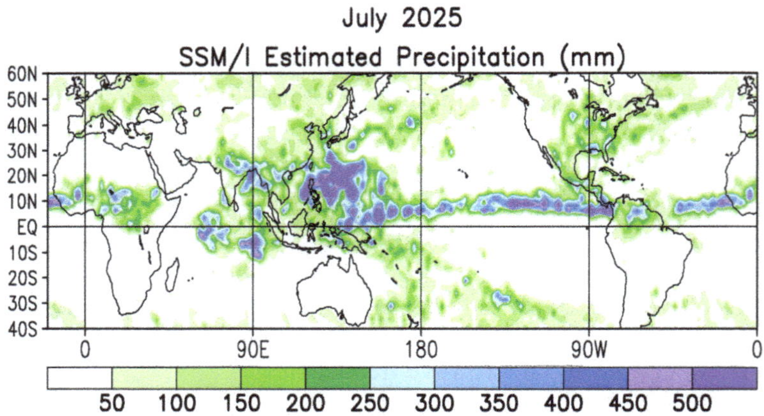

Figure 3.3.2.1 Monthly rainfall (mm) estimated from DMSP SSM/I microwave data.
(Source: https://www.cpc.ncep.noaa.gov/)

Ferraro et al (1997) have explained how microwave data can be used for climatological applications. Figure 3.3.2.1 shows a map of global precipitation in mm for the month of July 2025 which shows highest values over the ITCZ and the Indian subcontinent typical of a monsoon situation.

3.4 Vertical Profiles of Temperature and Water Vapour

Knowledge of the temperature and humidity distribution in the vertical is essential for obtaining a complete three-dimensional representation of the atmosphere. The conventional method of measuring vertical profiles of temperature, water vapour or any other atmospheric parameter is to release a balloon carrying a suitable instrument package that would sense the parameter and transmit the data to a ground station. In a radiosonde, which is the most widely used instrument for this purpose, an electronic package with a transmitter is attached to the balloon and the sensors measure pressure, temperature and humidity at various levels as the balloon rises. The signals received at a ground station are decoded to obtain the vertical profiles of temperature, humidity and other parameters. Most upper air stations take two radiosonde ascents per day at the synoptic hours of 0000 and 1200 UTC.

Radiosonde data also have their own inherent errors due to solar radiation during daytime and infrared cooling at night. The time lag between the temperature and humidity measurements and errors in pressure measurement add further to the overall errors and uncertainties (McMillin et al 1988). In addition, national meteorological services use radiosonde instruments of different makes and designs, and employ different ground computation procedures. This requires them to undertake rigorous inter-comparisons periodically so as to maintain compatibility amongst the instruments and uniformity in the global synoptic analyses.

Over the years, however, these limitations have been more than overcome, as radiosonde and rawinsonde observations are now supplemented by satellite soundings which are repetitive, globally consistent and available over land and oceanic areas alike. Khanna and Kelkar (1993) have described the early attempts at retrieving the temperature soundings of the atmosphere over the Indian region by processing NOAA satellite data received at New Delhi with the International TOVS Processing Package. They have also explained the basic principles of the method adopted for the retrieval, which are summarized below.

The narrow band radiance measurements made by a satellite depend upon the surface skin temperature and the vertical profiles of temperature and absorbing gases. In the simplest case, it can be assumed that there are no clouds in the atmosphere, there is no scattering, the viewing geometry is vertical and the earth's surface is a black body. The outgoing radiance R_ν at frequency ν reaching the satellite can then be expressed as a sum of two terms:

$$R_\nu = (I_0)_\nu \tau_\nu(z_0) + \int_{z_0}^{\infty} B_\nu\{T(z)\} \frac{d\tau_\nu(z)}{dz}\, dz$$

In this Radiative Transfer Equation (RTE), the first term on the right hand side is the spectral radiance I_0 at frequency ν emitted by the earth's surface, denoted by z_0, and attenuated by the atmosphere. The second term is the integrated spectral radiance at frequency ν emitted by various layers of the atmosphere and attenuated by the atmosphere above them. B_ν is the Planck function for frequency ν and temperature T at height z, and τ_ν is the transmittance from the level z to the top of the atmosphere.

R_v can be regarded as a sum of the radiance emanating from the earth's surface and the contributions to the upward flux from various atmospheric layers. If we know the vertical profiles of temperature and optical depths of the absorbing gases, and the absorption and emission characteristics of the atmosphere, we can compute the value of the radiance R_v. This can be accomplished in a relatively straightforward manner and is aptly called the forward problem.

On the other hand, if we measure R_v at several frequencies in an absorption band, it should be possible to compute either the vertical profiles of temperature or the optical depths of atmospheric gases or both. This is called the inverse problem and it is a much more complicated process. The functions multiplying B_v in the integral term of the RTE indicate the layer which contributes the maximum to the radiance sensed by a given satellite channel. However, for each channel a considerable depth of the atmosphere contributes to the radiance. Further, satellite instruments sense radiation over a small range of frequencies rather than monochromatic radiation. It is therefore necessary to perform an integration over frequency to obtain radiances of adequate accuracy as measured by the satellite instrument. The inverse problem is again complicated by the fact that the transmittances are not known as a function of the optical depth and temperature as accurately as required. It is not, therefore, possible to infer both temperature and humidity profiles from the same set of satellite measurements. In order to obtain accurate temperature soundings it is necessary to make simultaneous measurements in the water vapour and ozone absorption bands. It is also desirable to make additional measurements in the short wave spectral region.

Radiance measurements in the absorption bands of CO_2 centred at 15 and 4.3 μ, water vapour at 6.7 μ and ozone at 9.6 μ provide the capability to derive temperature and moisture profiles and total ozone respectively. Sampling at the centre of the band yields radiation from the upper levels of the atmosphere, as radiation from below has already been absorbed by the atmospheric gas. Sampling away from the centre of the absorption band yields radiation from successively lower levels of the atmosphere. In the wings of the absorption band are the windows that view the bottom of the atmosphere.

The RTE can be written in a different form as

$$R_v = (I_0)_v \tau_v(z_0) + \int_{z_0}^{\infty} B_v\{T(z)\} K_v(z)\, dz$$

where K is what is called a weighting function. The normalized weighting functions for NOAA TOVS/HIRS channels in the CO_2 band are shown in Figure 3.4.1.

A weighting function is used to multiply or weight the Planck function in the atmospheric component of the emitted radiation. It indicates the relative contribution from a given level of the atmosphere to the radiance received by the satellite through a given channel. In order to understand better the nature of the weighting functions, we may consider three air parcels at different heights. For the lowest parcel, the atmospheric density is high and so the amount of radiation emitted is high, but almost all of it is absorbed by the atmosphere above and only a small residual amount reaches space. For the highest parcel, the transmittance to space is high, but comparatively little radiation is emitted because atmospheric density is very low at that height. The result of these opposing effects is that, at some intermediate height, the contribution of a parcel to the radiance is maximum. Most of the radiance originates in a layer around the peak of the weighting function. From a knowledge of the vertical distribution of atmospheric constituents and their spectroscopic properties measured in the laboratory, the variation of the weighting function with height can be calculated. Since the weighting function is the derivative of the transmittance profile, it will peak higher in the atmosphere for the frequency at which the absorption is stronger. In this way, a carefully selected family of frequencies will be able to sense radiation from different layers in the atmosphere.

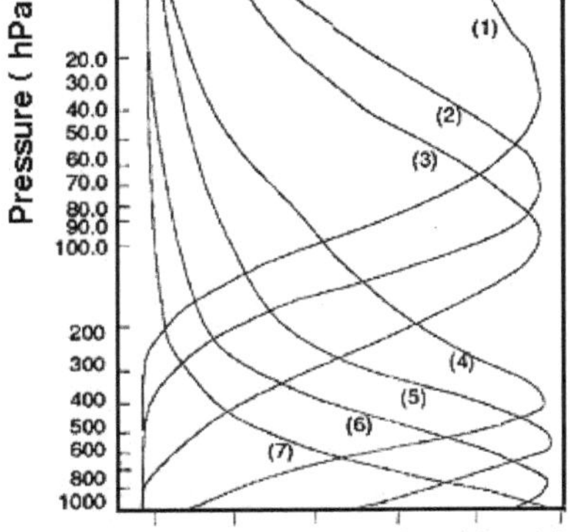

Figure 3.4.1 Normalized weighting functions for NOAA TOVS/HIRS channels in the CO_2 band.

Many problems are, however, encountered when dealing with weighting functions. Firstly, weighting functions do not have sharp peaks, so the radiance can be attributed only to the mean properties of thick layers and not to those of single levels. Secondly, successive weighting functions have a high degree of overlap, so although the instrument may make measurements at N separate frequencies, we obtain fewer than N pieces of independent information. Both these factors have implications in the inversion procedure.

The basic problem, however, is that we are trying to retrieve T(z) from a finite number of measurements. T(z), however, is a continuous function of height z and ideally requires an infinite number of parameters to represent it fully. This means that theoretically speaking, an infinite number of atmospheric profiles can lead to the same radiances being measured by the satellite. The aim of a good retrieval scheme is to find out the best or the most reasonable profile.

There are three types of retrieval algorithms, statistical, physical and hybrid, none of which can be said to have a clear superiority over the others. The statistical retrieval method is the simplest as it does not involve the RTE at all. This approach uses a statistical knowledge of the properties of the atmosphere. A large data set of observations from radiosonde ascents and satellite radiances, co-located as closely as possible in time and space, is first compiled. These are subjected to a regression analysis and a set of correlation coefficients is generated. The regression relationship is then applied to the actual radiances to get the vertical temperature profiles. The procedure is simple and requires minimum computer power and time. In fact it does not make use of either the RTE or the transmittance functions. However, the large training data set needs to be periodically updated, say once a week, to account for any gradual changes in the channel characteristics or air mass composition. As the satellite viewing geometry may vary during different passes over the same area, the radiance may have to pass through varying depths of the atmosphere, in contrast to near vertical radiosonde ascents, and hence a limb correction has to be applied. Some error will also arise if the emissivity of land, which depends upon soil moisture, is not parameterized realistically.

Physical retrieval algorithms attempt to solve the RTE in just one step in perturbation form or iteratively in a number of steps. In the one-step method, the RTE is solved to get simultaneous temperature and moisture profiles using all available radiation information. The iterative process starts with a first guess temperature profile which is judiciously chosen from a model prediction or climatology. The weighting functions are calculated and the RTE is solved to get estimates of the radiance in each channel. If the computed radiances match the observed ones within an acceptable margin, the current profile is accepted as the solution. If the computation diverges, the current profile is adjusted and the computations are repeated until a convergence is achieved.

This is a good approach because the physical processes are clearly evident at each stage. However, the method is computationally intensive and it requires an accurate knowledge about the transmittances, which have to be tuned for view angle, cloudiness and air mass variations.

Physico-statistical or hybrid retrieval algorithms are a compromise between the two approaches. They use the inversion technique to make retrievals but the initial guess is made using the statistical approach. They do not use a large training data set. They also use weighting functions like in physical retrievals, but they do not directly perform an integration of the RTE. The 3I method (Improved Initialization Inversion) is based upon the TIGR (Thermodynamic Initial Guess Retrieval) data set which is generated as a one-time effort. The nearest profile can be chosen from this voluminous set of 80,000 radiosonde ascents, made over 2,000 locations in tropical as well as extra-tropical regions covering land and sea at different viewing angles.

Khanna et al (1993) found that satellite retrievals over the Indian region were in agreement with radiosonde-measured temperatures between 700 and 150 hPa. Large differences were noticed, however, below 850 hPa and above 150 hPa levels due to excessively high surface temperatures in summer and large seasonal variations in atmospheric moisture. There were particular problems while dealing with thick persistent cloudiness during the monsoon season.

3.4.1 NOAA TOVS

The NOAA-6 satellite first carried an instrument called the TIROS Operational Vertical Sounder (TOVS) and satellites up to NOAA-14 continued to carry the same (Smith et al 1979). TOVS consisted of three separate parts: High Resolution Infra-red Radiation Sounder Version 2 (HIRS/2), Microwave Sounding Unit (MSU) and Stratospheric Sounding Unit (SSU). The system as a whole provided atmospheric temperature profiles from the surface to 10 hPa, water vapour content at three levels of the atmosphere and total ozone content. The HIRS/2 was a step-scanner 20-channel spectrometer with 19 IR channels and one VIS channel.

Later satellites of the NOAA series from NOAA-15 onwards flew a higher version of TOVS called the Advanced TOVS (ATOVS) system with advanced versions of all its subsystems. HIRS/3 was a discrete stepping, line-scan instrument designed to measure scene radiances in 20 spectral bands to permit the calculation of the vertical temperature profile from the earth's surface to about 40 km. Multi-spectral data are obtained from a single telescope and a rotating filter wheel containing 20 individual filters from 1 VIS channel (0.69 μ), 7 SWIR channels in the CO_2 band (4.57, 4.52, 4.47, 4.45, 4.13, 4.00, 3.76 μ), 7 TIR channels in the CO_2 band (14.95, 14.71, 14.49, 14.22, 13.97, 13.64,

13.35 μ) and 5 TIR channels in the H_2O and ozone bands (12.47, 11.11, 9.71, 7.33, 6.52 μ).

HIRS/3 had an elliptical scan mirror which provided cross-track scanning in 56 increments of 1.8° each. The mirror stepped rapidly in less than 35 msec, then held at each position for 100 msec for the 20 filter segments to be sampled. The IFOV for each channel was approximately 1.4° in VIS and SWIR and 1.3° in the TIR band which, from an altitude of 833 km, covered a ground area of 20.3 km × 18.9 km at nadir.

3.4.2 NOAA AMSU

Beginning with NOAA-15, all subsequent NOAA satellites have flown the 20-channel Advanced Microwave Sounding Unit (AMSU) which consists of two parts: 15-channel AMSU-A for temperature and moisture retrievals and 5-channel AMSU-B for moisture retrieval. AMSU weighting functions spread out much better than the earlier instruments. AMSU-A is designed to measure not only the vertical temperature profiles but to derive many atmospheric and surface parameters, such as precipitable water vapour, rain rate, cloud liquid water, sea ice and snow cover. AMSU-A is a cross-track, line-scan instrument designed to measure scene radiances in 15 discrete frequency channels. There are three channels at 24, 31 and 89 GHz, and 12 channels between 50 and 57 GHz. These measurements permit the calculation of the vertical temperature profile from the surface to about 3 hPa or 45 km height. The ground resolution is 48 km at nadir and the swath width is 2,343 km from the 833 km nominal orbital altitude.

AMSU-B is another cross-track, line-scan instrument that measures scene radiances in 5 channels. Of these two are at 89 and 150 GHz, and three near 183 GHz. The AMSU-B channels are numbered 16 to 20, following the 15 channels of AMSU-A. Channels 18, 19 and 20 span the strongly opaque water vapour absorption line at 183 GHz to provide data on atmospheric humidity. Channels 16 and 17 penetrate through the atmosphere to the earth's surface. The ground resolution is 16 km at nadir which is much higher than that of AMSU-A.

3.4.3 GOES Sounders

Although the physical principle of atmospheric sounding is the same for a polar orbiting or geostationary satellite, sounding instruments on geostationary satellites demand far greater sensitivity and precision because of the small amount of radiances that they have to measure at geostationary height in the same spectral bands. GOES was the first geostationary meteorological satellite to carrying a sounder. The first GOES sounder was the VISSR Atmospheric Sounder (VAS) which was rather a more advanced version of the Visible Infrared Spin Scan Radiometer (VISSR) carried by earlier GOES satellites.

In VAS, west-east scan lines were generated by the rotation of the spinning spacecraft. The lines were moved by stepping the mirror to scan north-south. The resolution was 13.8 km for channels. For atmospheric sounding, VAS had a filter wheel for measuring radiances in 7 bands in the region of CO_2 absorption, 3 in the region of water vapour absorption, and two in the IR window regions.

The GOES I-M sounder, launched on the 3-axes stabilized GOES satellites between 1994 and 2001, was a scanning radiometer with filter wheel selection. It had 18 IR bands plus a low-resolution VIS band, some of which had never been used before in geosynchronous orbit and were sensitive to temperature, moisture and ozone. While VAS had to share scanning time with the imager, the GOES I-M sounder worked independently, enabling operational soundings with a resolution of 10 km between 60 ºN and 60 ºS latitudes. The area and frequency of scan were fully programmable. The scanning mirror stepped in 10 km increments, with a step and dwell time of 0.1 sec. Initially, the GOES I-M sounder spectral selection was patterned after the NOAA HIRS sounder channels but subsequently many modifications were introduced.

The GOES-N satellite was launched in May 2006. The GOES N-P Sounder was a 19-channel discrete-filter radiometer covering the spectral range from the visible channel wavelengths to 15 μ. It has 7 channels in TIR, 5 in MWIR, 6 in SWIR and 1 VIS channel for which there were four sets of detectors. Approximately one-fourth of the wheel has no filters. While this dead zone was rotating in front of the detectors, the scanning mirror was stepped to the next location then stopped while the channels were sampled. The mirror scanning sequence took place by sweeping in alternate east-west and west-east directions, similar to the imager.

3.4.4 DMSP SSM/T

The early DMSP satellites carried, in addition to the SSM/I imager, a temperature profiler called SSM/T. This was a 7-channel microwave sounder that measured atmospheric emission in the 50-60 GHz oxygen band. The SSM/T was a cross-track scanning radiometer having a field of view of 14.4°. There were 7 total cross-track scan positions with an angular separation of 12° and with a maximum cross-track scan angle of 36°. While the ground resolution at nadir was about 174 km, it degraded at the far ends of each scan to 213 x 304 km. The SSM/T data swath was about 1500 km and there was a data coverage gap between successive orbits.

One more payload that DMSP satellites initially carried was the SSM/T-2 water vapour profiler, which was a 5-channel cross-track scanning microwave radiometer. Three out of the five channels were situated symmetrically about the 183.31 GHz water vapour resonance line and the remaining two were window channels. SSM/T-2 was designed to provide global monitoring of the concentration of water vapour in the atmosphere under all sky conditions by taking advantage of the reduced sensitivity of the microwave region to cloud attenuation.

The SSM/T and SSM/T-2 scan mechanisms were synchronized to make their measurements coincident. The swaths had the same width of 1500 km. The SSM/T-2 completed 7.5 scans per minute. There were 28 observations or beam positions per scan for each of the five channels, with each observation having a spatial resolution of about 48 km.

SSM/T-2 data were found to provide excellent signatures of mid-latitude weather phenomena such as fronts and extra-tropical cyclones. Other phenomena such as tropical cyclones, and the extent of sea ice and snow cover could also be identified. The retrieval of vertically integrated water vapor was also possible due to the strong sensitivity of the 183.31 GHz water vapour absorption line.

3.4.5 INSAT-3D Sounder

The INSAT-3D satellite launched in 2013, carried a sounder payload which has made it much easier to get the data in real time and process it faster, enabling its quick assimilation by numerical models. At that time INSAT-3D was the only geostationary satellite besides GOES, to have a sounding capability (Hawkins et al 2008). The instrument could make soundings at 10 km ground resolution every 3 hours for a full frame scan. Derivation of vertical profiles of temperature and humidity over 30 x 30 km areas was possible. Besides these vertical profiles, there were many other parameters than could be derived such as atmospheric stability indices, total precipitable water, and total column ozone. The characteristics of the 19 channels of the INSAT-3D sounder are given in Table 3.4.5.1. The INSAT-3DR and -3DS satellites that are currently operational, carry identical instruments.

Table 3.4.5.1 Characteristics of INSAT-3D/3DR/3DS Sounder Channels.

Channel Number	Central Wavelength (μ)	Main Absorbing Gas
1	14.71	CO_2
2	14.37	CO_2
3	14.06	CO_2
4	13.96	CO_2
5	13.37	CO_2
6	12.66	H_2O
7	12.02	H_2O
8	11.03	Window
9	9.71	O_3

Table 3.4.5.1 *Contd...*

Channel Number	Central Wavelength (μ)	Main Absorbing Gas
10	7.43	H_2O
11	7.02	H_2O
12	6.51	H_2O
13	4.57	N_2O
14	4.52	N_2O
15	4.45	CO_2
16	4.13	CO_2
17	3.98	Window
18	3.74	Window
19	0.695	Visible

The INSAT-3D sounder channels have a commonality with those of the GOES-12 sounder and the High Resolution Infra-red Radiation Sounder (HIRS) which is a component of the NOAA ATOVS. The retrieval algorithm for the INSAT-3D sounder is therefore similar to the HIRS and GOES algorithms developed by Cooperative Institute for Meteorological Satellite Studies (CIMSS), University of Wisconsin. The algorithm includes generation of the hybrid first guess atmospheric profiles using a linear combination of regression retrieval and NWP model forecast. This is followed by a non-linear physical retrieval procedure to make the first guess consistent with the sounder observations. Details of the retrieval process are documented by IMD (2014).

The INSAT-3D/3DR/3DS retrievals can be made at 10 or 30 km resolution at 30 vertical pressure levels up to 0.1 hPa for temperature and at 21 levels up to 100 hPa for water vapour. Other sounder-derived products include geopotential height, layer and total precipitable water, lifted index, dry microburst index and maximum "vertical" θ_e differential.

The sounding retrievals are useful for nowcasting and short-range predictions of severe weather events. Gopikrishnan et al (2023) made a study of collocated measurements of INSAT-3D and INSAT-3DR with those from 18 radiosonde stations, Atmospheric Infrared Sounder (AIRS) retrievals, and reanalysis data within the troposphere to validate the INSAT soundings for India land regions. Their validation showed that temperature profiles are generally accurate within 1–1.5 °K, Nevertheless, a better cloud filter together with an advanced radiative transfer model can further improve the algorithm used for retrieval. Similar validation studies have been carried out by Rao et (2020) and Giri et al (2022).

3.5 Sea Surface Temperature

It was known since long that oceans influence weather patterns over large regions and that they have close linkages to phenomena like the monsoons. It was a century ago that Sir Gilbert Walker, after whom the Walker circulation has been named, had discovered three large oscillations in which pressure in one region varies inversely with that in another region. These were the North Atlantic Oscillation, the North Pacific Oscillation and the Southern Oscillation (Walker 1923, Kelkar 2015). However, the real impetus to monitoring the El Nino and studying its global impacts came only in the late 1980s, when a reliable time series of global sea surface temperatures derived from NOAA satellites became available. For much of what we know today about the El Nino phenomenon, the genesis of El Nino and La Nina events, their intensity, evolution, spread and retreat, ocean-atmosphere interactions, and El Nino's impacts on the global climate, due credit has to be given to the extensive as well as intensive monitoring of SST that has been done by satellites (Figure 7.4.2). Long-term monitoring of the oceans is now considered essential for detecting signals of climate change and for understanding the processes that trigger and maintain inter-annual and longer time scale variability of climate. Ocean observations are equally important for weather forecasting and other practical applications like in marine industries and fisheries.

3.5.1 Properties of Ocean Water

Temperature, salinity and density are the three basic characteristics of ocean water. However, unlike the atmosphere, the ocean is horizontally stratified, as these basic ocean characteristics vary much more with the ocean depth than in the horizontal direction. The temperature may drop by 5 °C from the sea surface to a depth of 1 km, but the horizontal temperature gradient may be of the order of only 5 °C in 5000 km. The annual mean temperature of the global oceans is 3.5 °C.

Sea water is a complex solution of positively and negatively charged ions, and non-ionized organic and inorganic compounds. Although the concentration of the dissolved salts varies from place to place, the percentage of abundant elements remains almost a constant. The total quantity of solid material contained in sea water when all the material is completely oxidized, is called its salinity and is expressed in ppt (parts per thousand) or as a percentage. The average salinity of sea water is 35 ppt or 3.5 %. As very small changes in salinity can significantly alter the density of sea water, salinity has to be measured with an extremely high accuracy. The most precise and widely used method for salinity determination is by measurement of its electrical conductivity. Salinity is expressed as the ratio of the electrical conductivity of sea water to the electrical conductivity of a solution of potassium chloride of a standard concentration.

The distribution of salinity at the sea surface is again basically zonal in nature. Average surface salinity is lowest just north of the equator and highest in the sub-tropical regions. It again decreases rapidly at higher latitudes. Higher values of salinity are observed where evaporation exceeds precipitation. Lower values of salinity are found near the coastlines, in regions of fresh water influx from rivers, and also in the polar regions where ice freezing discards salt to lower waters and melting leads to addition of fresh water.

The vertical profile of salinity exhibits a marked minimum at depths between 600 m and 1 km, and then an increase to about 2 km depth. In the tropics, a sharp maximum is found at around 100 m depth just above the thermocline. In high latitudes, the surface value of salinity is low but it increases with depth to about 2 km and thereafter assumes a constant value.

The density of ocean water is an important parameter as many oceanic movements are driven by differences in density. The density ranges from about 1021 kg m^{-3} at the surface to 1070 kg m^{-3} at 10 km depth and it is a function of salinity, temperature and pressure.

3.5.2 The Layered Thermal Structure of the Ocean

Most of the solar radiation incident on the ocean is absorbed very close to the surface. Wind blowing on the ocean stirs the upper layers, producing a warm thin layer at the top of the ocean having a uniform temperature and salinity. This upper layer of the sea extending from the surface down to a depth of about 10 to 200 m is called the mixed layer. The sea water below about 1 km depth also has a uniformly cool temperature and it is called the deep layer. The region of transition between the mixed layer and this deep layer is called the thermocline and it is characterized by a steep temperature gradient.

The thermocline is also defined as the depth at which the temperature gradient is maximum. However, it is difficult to mark the depth of the thermocline very precisely. In low and middle latitudes the thermocline is always present at depths between 200 and 1000 m, but at higher latitudes where the SSTs are lower, the thermocline is present only seasonally.

The distribution of temperature and salinity at the ocean's surface is influenced by several factors such as heat fluxes, rate of evaporation, rainfall, fresh water inflows, and freezing and melting of sea ice. The mixed layer tends to be more saline than the thermocline between 10° and 40° latitudes, where evaporation exceeds precipitation. At higher latitudes, the mixed layer is less saline because of rain and melting ice.

3.5.3 Remote Sensing of Ocean Parameters

In situ measurements of ocean parameters are very important but it is extremely difficult to cover the vast expanse of the oceans, that too in three dimensions. The cost of operating and maintaining meteorological and oceanographic research vessels is high, and there are only a few of such dedicated vessels. There is a voluntary observing fleet of merchant navy ships which undertake to carry out the additional responsibility of reporting weather observations and keeping weather logs, but a major constraint in this effort is that such observations are confined to narrow designated shipping lanes in which the merchant vessels sail. As a result, vast areas of oceans still continue to remain uncharted.

Against the backdrop of these difficulties and limitations, satellite remote sensing now provides us with the capability to monitor four important parameters over the global oceans (Pearce et al 2001):

(a) Sea Surface Temperature (SST), which has certainly been most extensively mapped by satellite remote sensing techniques, shows up ocean circulation features, thermal fronts, eddies and upwelling zones and is a basic tool for monitoring the El Nino phenomenon,

(b) Ocean colour, which is an indicator of the presence of phytoplanktons, particulates and suspended sediments in the top layer and surface foam, reveals the properties of the seabed, and shows ocean circulation patterns, fronts and upwelling regions,

(c) Sea surface elevation, which is a product of satellite altimetry, helps to study the shape of the marine geoid, and provides information on tides, and

(d) Sea surface roughness, which is a measure of the effect of wind on the sea surface, and is estimated by scatterometers and synthetic aperture radars.

All the four ocean variables have a direct bearing on both oceanic and atmospheric processes.

3.5.4 Satellite-derived Sea Surface Temperature

Sea Surface Temperature (SST) is an important geophysical parameter, providing the boundary condition used in the estimation of heat flux at the air-sea interface. On the global scale, it is important for building global climatologies, in weather and climate modelling, in studies of the earth's heat balance and for gaining an insight into atmospheric and oceanic circulation patterns such as the El Nino. On a more local scale, SST can be used operationally to assess eddies, fronts and upwellings for marine navigation, detection of cyclogenesis and measuring biological productivity.

A satellite radiometer measures radiation over finite wavelength bands. The radiance measured has an associated brightness temperature, or the temperature at which a black body would emit the same radiation. From this the body's true surface temperature can be computed if the emissivity is known. As the emissivity of the sea surface can be assumed to have a value of 1, which is a very close approximation, the satellite-measured TIR window radiance can be inverted to get the temperature of the sea surface. This is the basic principle of satellite SST retrievals.

It is important to consider what exactly is meant by the sea surface. Because of the existence of vertical temperature gradients in the upper layer of the ocean, the SST will be different depending upon the depth at which measurements are made. If we imagine the sea surface to have a thickness of just of one or two molecules, we should ideally measure an interface SST, but it cannot be done in actual practice. Just below this, at a depth of 10 μ, we will get what is called the skin SST. This is what a satellite measures in the TIR channel. The sub-skin SST is at a depth of 1 mm and is what a satellite measures in the microwave channel. A measurement further down is commonly referred to as the bulk SST or near-surface SST. This is measured by a ship or buoy. The different SSTs will have different values at any time or place. The diurnal temperature cycle will cause these differences to be greatest during the afternoon and smallest just before sunrise. Comparisons of SSTs measured by ships or buoys with satellite SSTs are, therefore, not a straightforward matter. Effects of depth, diurnal heating and evaporative cooling may lead to differences as large as 1 °C. These differences have to be properly parameterized when blended SST data sets are compiled from various data sources.

The 3.55-3.93 μ channel has the advantage that it is less sensitive to atmospheric water vapour. However, it admits a substantial amount of reflected solar radiation. It is, therefore, primarily used at night. The 10.3-11.3 μ and 11.5-12.5 μ channels are more affected by water vapour, but are not substantially contaminated by reflected solar radiation. It is the judicious combination of radiance measurements from channels 3, 4 and 5 that permits extraction of SST.

There are two major limitations of satellite SST retrievals. Firstly, clouds block infrared radiation from the earth's surface from reaching the satellite and if cloudy regions are not appropriately identified, SST retrievals would turn out to be incorrect. A cloud filtering scheme is therefore required. Secondly, there is a residual attenuation of the emission from the sea surface on its upward path and scattering by aerosols and atmospheric water vapour. This has to be compensated by applying an atmospheric correction. The correction is a function of location and season and may be as large as 4-8 °C.

These problems can be largely circumvented by adopting certain procedures, but not totally eliminated. Maps of SST compiled from TIR measurements are often made as weekly or monthly composites which allow enough time to capture cloudfree pixels over a region. The effects of the intervening atmosphere, being wavelength-dependent can be further corrected for by resorting to multi-channel measurements. This was not possible with the VHRR instruments of INSAT-1, INSAT-2 and Kalpana satellites which had only one TIR channel.

Considering that two different window channels will have different attenuations and therefore different brightness temperatures, for nadir measurements in cloudless regions the sea surface temperature, T_b, should be of the form

$$T_b = a_0 T_i + a_1 (T_i - T_j) + a_2$$

where T_i and T_j are the brightness temperatures determined from the radiance values in two different infra-red window channels i and j. The constant a_0 is close to 1 suggesting that the infra-red temperature measured in any one of these channels is close to the sea surface temperature. The temperature difference term makes a small correction to this temperature for atmospheric transmittance. The a_2 term is another small correction factor associated with the different atmospheric brightness temperatures at different channels.

In general, there are three different classes of SST derivation algorithms.

(a) The split-window algorithm uses the T_{11} brightness temperature as the lowest order estimate of sea surface temperature and the difference $T_{11} - T_{12}$ to correct for the atmosphere.

(b) The dual-window algorithm uses the T_{11} brightness and the difference $T_{3.7} - T_{11}$ to correct for the atmosphere.

(c) The triple-window algorithm uses the T_{11} brightness and the $T_{3.7} - T_{12}$ difference to correct for the atmosphere.

In addition, there are correction terms that must be applied to adjust the measurements made off nadir. If we define θ to be the sensor zenith angle, then the three algorithms have the form:

(Night-time - Triple)

$$SST_{3/4/5} = a_0 T_{11} + a_1 (T_{3.7} - T_{12}) + a_2 (T_{3.7} - T_{12}) (\sec\theta - 1) + a_3 \sec\theta + a_4$$

(Night-time - Dual)

$$SST_{3/4} = a_0 T_{11} + a_1 (T_{3.7} - T_{11}) + a_2 (T_{3.7} - T_{11}) (\sec\theta - 1) + a_3 \sec\theta + a_4$$

(Daytime - Split)

$$SST_{4/5} = a_0 T_{11} + a_1 (T_{11} - T_{12}) + a_2 (T_{11} - T_{12}) (\sec\theta - 1) + a_3 \sec\theta + a_4$$

The retrieval process involves three steps: (1) Exclusion of data at large zenith angles since experience has shown that the SST retrievals degrade as the sensor zenith angles increases, (2) Finding cloud-free pixels for which there are several cloud detection techniques, (3) Applying an appropriate SST extraction algorithm.

Since the INSAT-3D satellite has a 6-channel imager, it is now possible to derive SSTs from the split thermal window channels (10.2-11.3 μ, 11.5-12.5 μ) during daytime and the mid IR window channel (3.7-4.1μ) during night-time over cloud free oceanic regions (Figure 3.5.4.1). The required atmospheric correction has been made through a suitable characterization of the tropical atmosphere in radiative transfer model to determine the brightness temperatures of INSAT-3D channels and then generating the regression coefficients for SST retrieval.

During daytime and for cloudfree pixels, INSAT-3D SSTs are derived using equations of the form

$$SSTs = A_0 + A_1 * T_{11} + A_2 * dT + A_3 * dT^2$$

where A_0, A_1, A_2, and A_3 are satellite zenith angle dependent regression coefficients, T_{11} and T_{12} are the brightness temperatures of the split-window channels and dT is the difference between them.

During night-time and for cloudfree pixels, INSAT-3D SSTs are computed as

$$SSTs = B_0 + B_1 * T_3 + B_2 * dT + B_3 * dT^2$$

where B_0, B_1, B_2, and B_3 are satellite zenith angle dependent regression coefficients and T_3 is the brightness temperature for channel 3.

The single channel INSAT-3D SST is computed as

$$SSTs = a + b * Tb + c * WV$$

where Tb is the brightness temperature of the IR imager channel, WV is the total water vapour content, and a, b and c are regression coefficients (IMD 2014).

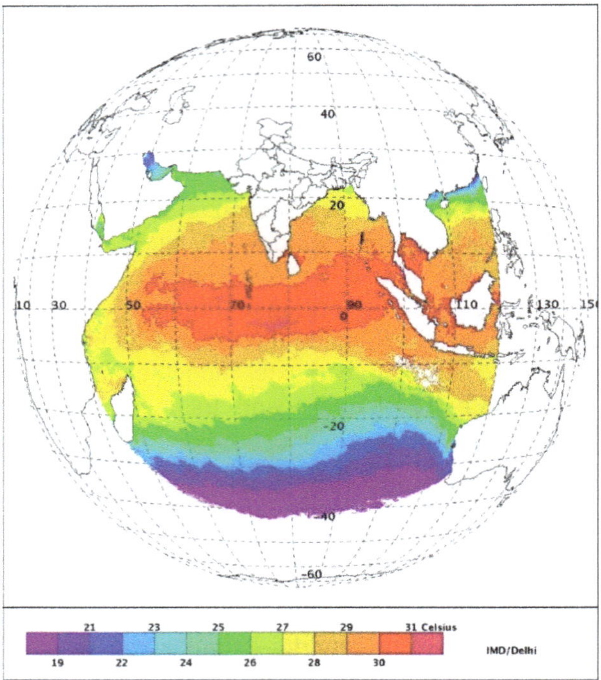

Figure 3.5.4.1 An example of a weekly composite SST map generated from INSAT-3D data for 24-30 April 2017. (Source: IMD)

3.5.5 Microwave SST Retrievals

SST retrievals from passive microwave measurements have the limitation of lower spatial resolution. This is due to the inherent problem of the low energy associated with radiation in the microwave region of the spectrum compared with that in the TIR region. However, microwave retrievals have an advantage over TIR in that radiation at these longer wavelengths is largely unaffected by clouds. As clouds and aerosols are essentially transparent to microwave radiation at frequencies below about 12 GHz, microwave remote sensing has the potential to eliminate the atmospheric contamination problems that vitiate TIR SST retrievals (Chelton et al 2005).

The atmospheric contribution to the microwave brightness temperatures measured by a satellite at these frequencies is only 2-5 per cent. Microwave estimates of SST are possible because the surface radiance is proportional to SST at frequencies between about 4 and 12 GHz. However, microwave signal returns are affected by wind-generated roughness at the ocean's surface and precipitation. The resulting changes in the surface emissivity can usually be accounted for by using multiple frequencies.

Since microwave emission from the sea surface is weak, accurate satellite microwave observations of SST require a highly calibrated and sensitive radiometer. High quality microwave SST data first became available in December 1997 from measurements at 10.7 GHz by the TRMM Microwave Imager TMI (Wentz et al 2000). Individual TMI observations of SST had an accuracy of 0.5 °C and a footprint of 46 km. This was very large compared with the 1 km resolution of satellite IR observations but the advantage was in that microwave SST retrievals were available in all non-raining conditions. Rain-contaminated measurements from TMI were identified and eliminated from further analysis using co-located measurements at 37 GHz. A limitation in coastal regions was that the TMI brightness temperatures were contaminated by land in the antenna side lobes within about 50 km of land. Another limitation of TMI was that its geographical coverage was limited to the latitude band of 40°S-40°N.

An attempt was made to eliminate all the limitations of the TMI mentioned above in the Advanced Microwave Scanning Radiometer (AMSR-E) on the NASA Aqua satellite. The AMSR-E covers 89% of the globe each day and 98% of it every two days. AMSR-E retrievals of SST are based on measurements of brightness temperature at 6.9 GHz, which is more sensitive to SST than the 10.7 GHz channel. The AMSR-E footprint size for SST measurements is 56 km, and the accuracy is estimated to be about 0.4°C

3.6 Potential Fishing Zones

An important example of how meteorological satellite data can be used for non-meteorological applications is the delineation of potential fishing zones in the oceans. Oceanic features like fronts, eddies, gyres and upwelling areas play a major role in the supply of nutrients in oceanic waters. Most of the productive fishing grounds of the world are located in the areas where upwelling is found to occur. Satellite-derived SSTs serve as a very useful indicator of prevailing and changing environmental conditions from the point of view of fish aggregation.

The Indian National Centre for Ocean Information Services (INCOIS), Hyderabad, makes extensive use of satellite-based estimates of chlorophyll concentrations and retrievals of SSTs for identification of Potential Fishing Zones (PFZ) in the Arabian Sea and Bay of Bengal. SST data are composited over 3-4 days and relative thermal gradients are worked out. Hence high SST gradients and high chlorophyll concentration are the potentially prospective areas for pelagic fish catch. The thermal front areas where fish are expected to congregate are delineated as PFZs. The PFZ advisories in the form of maps and text in local languages is then disseminated to about 600 fish landing centres located along the Indian coast daily through different media by

INCOIS. The information specifies the distance of the PFZ from a known coastal point, bearing and the depth of water at the zone. Adequate training is imparted to the fishermen in the utilization of the information including the guidance for reaching the PFZ location in the sea.

Solanki et al (2001) used near-synchronous SeaWiFS-derived chlorophyll concentrations and NOAA-AVHRR SSTs to understand patterns, persistence and inter-relationship between ocean colour and thermal features. An inverse relationship between chlorophyll concentration and SST features was observed. The frontal zone positions on the chlorophyll concentration images coincided with temperature boundaries at some locations. In a subsequent study by Solanki et al (2003), validation of the PFZ forecasts was carried out with the help of fishing trawlers plying in the PFZ regions.

In a later study by Nammalwar et al (2013) based on data from April 2007 to March 2011, the fish catch in the area covered by PFZ advisories was seen to be 3-4 times higher than that in the non-PFZ area along the coast of North Tamil Nadu. In a similar investigation around the Andaman and Nicobar Islands by George et al (2013) the fish catch was found to be significantly higher where PFZ advisories had been followed by fishing trawlers. Recently, Majumder et al (2021) have proposed a hybrid decision tree model developed for characterizing PFZ in terms of low, medium or high category of fish catch.

3.7 Ocean Colour

The colour of the sea is not a parameter that is of direct interest to meteorologists and atmospheric scientists. However, satellite observations of ocean colour have brought out the fact that it has a strong bearing on the structure of the thermocline. The penetration of the blue component deeper into the water and the absorption of the red component in the sea can alter the depth of the mixed layer and cause perturbations in the ocean circulation. Thus ocean colour has an indirect linkage with atmospheric processes.

The colour of the sea varies from deep blue to green or even greenish yellow due to scattering of short wave radiation by the water molecules. In clear waters, most of the red and yellow wavelengths in sunlight get absorbed in the upper few metres of sea water, and only the blue wavelengths are scattered, imparting an overall blue colour to the water. In the presence of particulate matter in the sea water, other wavelengths are also scattered, creating a different hue. When marine organisms called phytoplanktons are present in the water, the chlorophyll content absorbs the blue light and alters the colour to green. This is also seen in coastal waters where vegetation is dense and rivers bring in dissolved organic substances. Subtle changes in ocean colour can therefore be used as proxy indicators of chlorophyll concentration and primary production by phytoplanktons.

Phytoplankton cells contain dissolved CO_2 and therefore represent a key element of the carbon cycle in the ocean. They also control the optical turbidity of the oceans and the development of ocean ecosystems. Ocean colour information is important for the management of marine resources and protection of the coastal zone.

The Coastal Zone Colour Scanner (CZCS) instrument, which was launched on the Nimbus-7 satellite in October 1978, was the first such instrument that was solely dedicated to the observation of ocean colour and it worked until June 1986. The SeaStar (OV-2) satellite launched on 1 August 1997, carried an ocean colour sensor called SeaWiFS (Wide Field of View Sensor).

Table 3.7.1 Oceansat-1 OCM Spectral Bands.

Spectral Band Number	Wavelength Range (nm)
1	402-422
2	433-453
3	480-500
4	500-520
5	545-565
6	660-680
7	745-785
8	845-885

The Indian Remote Sensing Satellite IRS-P4 (named Oceansat-1), was launched in May 1999 from Sriharikota, using ISRO's Polar Satellite Launch Vehicle (PSLV). The satellite carried two payloads, the Ocean Colour Monitor (OCM) and a Multi-frequency Scanning Microwave Radiometer (MSMR). OCM was a multi-spectral CCD camera operating in eight narrow spectral bands in the VIS, NIR and IR range of the electromagnetic spectrum between 402 and 885 nm (Table 3.7.1). The camera was used to collect data on chlorophyll concentration, phytoplankton blooms, atmospheric aerosols and suspended sediments in the water. It had a spectral resolution of 360 m × 236 m and a swath of 1420 km. The satellite was in a polar sun-synchronous circular orbit at an altitude of 720 km with a repeat cycle of two days and an equator crossing time around noon. Oceansat-1 lasted for 11 years and was decommissioned in August 2010.

The successor satellite, Oceansat-2, was launched into a near-polar sun-synchronous orbit of 730 km altitude in September 2009. It carried three payloads: another OCM for the sake of continuity, a Wind Scatterometer called OSCAT and an Italian payload called ROSA which was a radio occultation sounder.

OCM data has been widely used for estimating chlorophyll concentrations and primary productivity of the ocean. In an interesting application of OCM data, Prasad et al (2000) used sequential data of IRS-P4 OCM to retrieve sea surface current velocities over parts of both east and west coasts of India. The method was based on matching suspended sediment dispersion patterns across two sequential atmospherically corrected and geo-referenced images by using the maximum cross-correlation technique.

The 36-channel Moderate Resolution Imaging Spectroradiometer (MODIS) on board NASA's Terra and Aqua satellites has been offering twice-daily coverage and simultaneous measurements of ocean colour, sea surface temperature and ocean primary production at a moderate spatial resolution of 250 m to 1 km. The Terra spacecraft has a 10:30 am equator crossing and the Aqua spacecraft has a 1:30 pm equator crossing time. Improved atmospheric products are enabling better correction for atmospheric effects in the ocean signals. The products are being used to investigate ocean circulation, ocean biology and ocean chemistry, including the ocean carbon cycle. In addition to ocean measurements, simultaneous measurements of land vegetation, cloud characteristics, temperature and moisture profiles, snow cover, and sea ice, are also being made.

3.8 Sea Surface Salinity

The global distribution of sea surface salinity was until recently derived from the limited measurements made by ships and buoys and it is quite different from that of sea surface temperature. High values of salinity are found near the centre of the ocean basins, away from the mouths of rivers, and in sub-tropical oceans, where evaporation exceeds precipitation. At high latitudes, salinity is low because of the melting of ice that dilutes sea water.

Fresh water is light and floats on the surface, while salty water is heavy and sinks. Greater salinity, like colder temperatures, results in an increase in ocean density with a corresponding depression of the sea surface height. In warmer, fresher waters, the density is lower resulting in an elevation of the sea surface. These height differences are related to the circulation of the ocean. The ocean stores more heat in its top 3 m layer than in the entire atmosphere. Salinity is therefore a key factor in the global thermohaline circulation and needs to be measured globally by satellite.

Microwave radiometry at low frequencies in the L-band (1.4 GHz, 21 cm wavelength) has come into use for estimating sea surface salinity. However, when deployed on a space platform, large antennas of the dimensions of several metres are required for achieving an adequate spatial resolution. To overcome this problem, the principle of aperture synthesis was adopted in the

Soil Moisture and Ocean Salinity (SMOS) satellite, which was launched by in 2009 as an ESA mission.

SMOS carries an instrument known as the Microwave Imaging Radiometer with Aperture Synthesis (MIRAS). This is a dual-polarized L-band microwave radiometer that has a ground resolution of 50 km and enables frequent and global coverage of these two important parameters. A two-dimensional interferometer with a set of elementary antennas that are uniformly distributed along a Y-shaped structure samples the Fourier transform of the scene. At the ground, an inverse Fourier transform is performed to retrieve the image. A complete description of the MIRAS payload has been given by McMullan et al (2008).

While SMOS is an ESA mission, there was another U.S.-Argentina joint mission named Aquarius for measurement of global sea surface salinity with a resolution of 150 km. However, it had a short life from August 2011 to June 2015. Aquarius carried an L-band polarimetric radiometer operating at a frequency of 1.413 GHz which was sensitive to salinity, and a polarimetric scatterometer operating at 1.26 GHz to correct for the ocean's surface roughness. It was in a 657 km sun-synchronous polar orbit with a repeat cycle of 7 days.

3.9 Soil Moisture

Soil moisture measurements on different spatial and temporal scales have many applications ranging from agricultural meteorology and operational hydrology to atmospheric modelling. Frequent *in situ* measurements of soil moisture on a global scale are extremely difficult because of the heterogeneity of soil and the time consuming nature of the traditional measurement process which involves drying and weighing soil samples. Hence data from satellite-based passive microwave radiometers have been used to make an indirect estimation of this important land parameter and to compensate for the absence of a good ground network (Rao et al 2001).

The relationship between the physical temperature of the surface T_s and its brightness temperature T_b measured by a satellite radiometer, is of the form $T_b = \varepsilon T_s$, where ε is the surface emissivity. In the microwave region, the emissivity is strongly dependent on the moisture content of the soil, varying between 0.6 for wet soils and 0.3 for dry soils. For a surface temperature value of 300 °K, the microwave brightness temperature may vary as much as between 180 and 270 °K for wet and dry soils respectively. This large range of variation of T_b provides a good signal-to-noise ratio and makes it possible to retrieve the soil moisture indirectly. Ideally, microwave estimates of soil moisture can be made with an accuracy as high as 99 %. However, in practice such an high accuracy cannot be attained because several other factors such as surface roughness, vegetation cover, or soil temperature, also influence the brightness temperature that is measured by the satellite.

While sea surface salinity and soil moisture are two apparently unconnected parameters, it so happens that both of them can be measured by a common satellite technique. The SMOS satellite described in the previous section could serve both purposes. The European Soil Moisture and Ocean Salinity (SMOS) mission was also a similar dual purpose mission. Thapliyal et al (2003, 2005) derived regression relationships between the Oceansat-1 MSMR 6.6 GHz channel data and soil moisture at various locations in India where *in situ* observations were available and found very good correlations between them.

An independent satellite for the measurement of soil moisture Soil Moisture Active Passive (SMAP) was launched by NASA in January 2015. While the active component is no longer functioning, the passive radiometer is still sending data. SMAP provides estimates of soil moisture in the top 5 cm of soil with an accuracy of the order of 0.04 cm^3/cm^3 at 10 km spatial resolution and 3-day average intervals over the global land area. It also provides estimates of surface binary freeze/thaw state in the region north of 45°N latitude which includes the boreal forest zone.

3.10 Vegetation Index

The first widespread application of meteorological satellite data for a non-meteorological purpose was perhaps in the derivation of what came to be known as the vegetation index. The pigment in plant leaves, chlorophyll, strongly absorbs sunlight (VIS 0.4-0.7 μ) and makes use of it in the process of photosynthesis. The cell structure of the leaves, on the other hand, strongly reflects radiation in the near infrared region (NIR 0.7-1.1 μ). Healthy plants with a high leaf density will absorb most of the incident visible light, and will reflect a large portion of the near infrared radiation. Plants which have not had a proper growth or have been struck by disease, will behave in an opposite manner, and reflect more of VIS and less of NIR radiation.

If a satellite radiometer has channels suitable for measuring radiation in VIS and NIR wavelengths and the view of the land is clear, the above properties of vegetation enable the derivation of a vegetation index that gives a gross idea of the state of the vegetation. There are many ways of defining a vegetation index, but the commonly used index is the Normalized Difference Vegetation Index (NDVI) expressed as

$$NDVI = (NIR - VIS)/(NIR + VIS)$$

where VIS and NIR are the radiances or brightness temperatures measured simultaneously for a given pixel in the VIS and NIR channels of a satellite radiometer.

Theoretically, NDVI can range between −1 and +1, with the value 0 signifying the total absence of vegetation, and an index of +1 indicating the maximum possible density of green leaves. The NDVI values derived in practice commonly lie between 0.1 and 0.7, the higher values of the index being associated with healthy vegetative cover. NDVI is best obtained under clear sky conditions, but this may not always be possible. In India, for example, the main agricultural season is during the monsoon when there is extensive cloud cover present most of the time. The presence of cloud and snow does vitiate the computations and may result in an under-estimate of the NDVI, creating the impression of a less green vegetation.

The net photosynthesis is directly related to the amount of photosynthetically active radiation (PAR) that a plant can absorb. If it absorbs more PAR during the growing season, the photosynthesis will be higher and the more productive it will be. If it absorbs less PAR, it will be less productive. These scenarios get reflected in the value of the NDVI which can be used to monitor the health and productivity of the vegetation and the incidence and spread of agricultural drought.

A variety of vegetation indices similar to the NDVI are being derived from different earth resources satellites using different channel combinations, but they have not become popular because of the long revisit intervals of such satellites. Hence the NDVI derived from NOAA AVHRR channel 1 and channel 2 data, also known as the Global Vegetative Index (GVI), has been most widely used on a global scale. Data sets of NDVI are available for computation of long-term averages and anomaly patterns.

In India, ISRO's National Remote Sensing Centre (NRSC) regularly produces NDVI maps using data from the Oceansat Ocean Color Monitor. Data from the Red and NIR bands are used and corrections are applied for sun-view angle to handle bidirectional reflectance effects. Atmospheric corrections are also applied using image-based techniques and cloud masking.

Prasad et al (2004) assessed the spatial distribution of NDVI for the period 1990-1999 over India and they found a good positive correlation between anomalies in rainfall and NDVI especially for monsoon season. They have prepared a climatic classification map of India based on NDVI and rainfall data and their temporal variation, and made an assessment of drought- affected regions. Milesi et al (2010) have analyzed the NDVI record for 1982-2006 together with climate, land use and crop production statistics to examine the trends in food grain production in India.

3.11 Snow Cover

Snow cover has many different effects on atmospheric and soil processes. Because of its high albedo, it reduces the fraction of solar radiation that can penetrate the soil and therefore has a cooling effect. Due to its small heat conductivity, snow cover acts as a thermal insulation between the atmosphere and the land surface. Melting snow serves as a heat sink as well as a source of many rivers such as those which originate in the Himalayas. That the Eurasian snow cover has a dominant and inverse relationship with the all-India summer monsoon rainfall, has been known since long, but this continues to be a topic of study (Bamzai et al 1999, Dash et al 2003). Conventional methods of measuring snowfall do not have the high accuracy or the spatial resolution required for quantifying such teleconnections. Therefore, remote sensing of snow cover assumes great importance for monsoon forecasting as well as climate studies (Kulkarni et al 2002).

Satellite-based estimation of snow cover is prone to errors because in VIS imagery snow and clouds both look alike and in IR imagery snow and land have the same temperature. Satellite mapping of snow cover is therefore done by using a suitable algorithm which takes the form:

$$\text{NDSI} = (\text{VIS} - \text{SWIR}) / (\text{VIS} + \text{SWIR})$$

where NDSI is a Normalized Difference Snow Index and VIS and SWIR are the reflectances measured in the respective channels. The computation is based upon the premise that snow reflects more in the VIS region of the spectrum than in the SWIR region. Moreover, since most types of clouds have a higher reflectance in the SWIR compared to snow, cloud and snow can be differentiated. The computation process also includes a preliminary cloud screening based on the temperature of the IR pixels (IMD 2014).

NASA monitors the earth's snow cover using the MODIS and VIIRS instruments on the Terra, Aqua, and Suomi NPP satellites, to create global snow cover maps. Figure 3.11.1 is a Terra MODIS view of the snowcapped mountains of the European Alps.

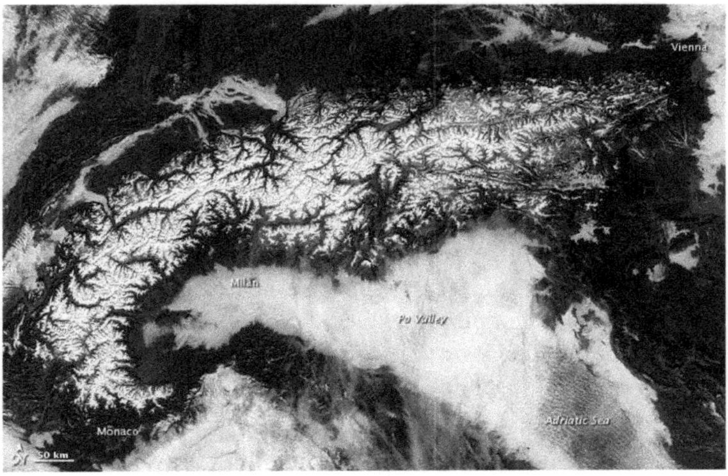

Figure 3.11.1 Terra MODIS image of 17 Jan 2011 showing snow and clouds over the European Alps.
(Source: https://earthobservatory.nasa.gov/images/)

3.12 Land Surface Temperature

Compared to SST, the land surface temperature could be said to be a poorly retrieved satellite parameter, with the main constraint lying in the large geographical variation in the properties of the land and soil surface. It is however required for many purposes such as in atmospheric models for the computation of sensible heat flux from the difference between the land surface temperature and the air temperature near the surface. SST is retrieved from satellite radiance measurements assuming an emissivity of 1, which is quite a valid assumption most of the time. In a similar manner, the Land Surface Temperature (LST) can, theoretically speaking, be derived from AVHRR radiances if the surface emissivity is known or assumed. However, unlike SST, LST has a high diurnal range and even if the emissivity problem is solved, it is another great challenge to interpolate the diurnal cycle from the twice daily values that can be obtained from AVHRR. Moreover, the need for correcting the retrieved temperature for atmospheric attenuation and cloud contamination applies to LST retrievals as well (Sobrino et al 1994).

The conventional screen temperature measured at a ground observatory, is not the soil temperature but the air temperature at the screen height. It is a thermodynamically defined quantity. LST, as may be measured by a radiometer mounted above the ground surface is, however, a radiatively defined quantity. Although these two parameters are quite related to each other, they differ in value and physical meaning. There are several other major problems that prevent NOAA AVHRR data from being used to develop a long-term LST data set, such as orbit drift, uncertainties in skin temperature retrieval due to unknown surface emissivity, unknown cloud occurrence during a day other than at observing time, or lack of LST diurnal cycle measurement.

Jin (2004) has made a serious attempt to overcome these problems and produce an LST Diurnal-averaged (LSTD) dataset of monthly skin temperature, monthly maximum skin temperature (T_{max}) and monthly minimum skin temperature (T_{min}), from 1981 to 1998 at 8 km resolution. The dataset is scaled up to a 0.5° × 0.5° latitude/longitude grid, and is comparable to other surface temperature data sets. LSTD can be used to study the climatology of land skin temperature and its diurnal, seasonal, and inter-annual variations on both global and regional scales. Snow screening is done through AVHRR Channel 1 and Channel 2 albedo information.

Global LSTD is useful because it spatially samples more areas than the global observation network and help to reduce any errors in the data-sparse regions. Another advantage of LSTD is that its uniformly high resolution makes it easier to study regional climate particularly over mountains, deserts or regions like the Tibetan plateau.

Li et al (2013) have made an exhaustive review of the current status and future perspectives of satellite-derived land surface temperature. They maintain that while LST is one of the key parameters in the physics of land surface processes on all scales, LST retrieval is still a challenging task since the LST retrieval problem itself is ill-posed. Different methods are available but they work in different conditions under limiting assumptions and are not universally applicable. The LST as a parameter is also not uniquely defined and its validation becomes all the more difficult.

3.13 References

Adler R. F. and Mack R. A., 1988, "Thunderstorm cloud height - rainfall rate relations for use with satellite rainfall estimation techniques", *J. Climatol. Appl. Meteor.*, 23, 280-296.

Arkin P. A. and Meisner B. N., 1987, "The relationship between large-scale convective rainfall and cold cloud cover over the western hemisphere during 1982-84", *Mon. Wea. Rev.*, 115, 51-74.

Arkin P. A., Rao A. V. R. K. and Kelkar R. R., 1989, "Large-scale precipitation and outgoing longwave radiation from INSAT-1B during the 1986 southwest monsoon season", *J. Climate*, 2, 619-628.

Bamzai A. and Shukla J., 1999, "Relation between Eurasian snow cover, snow depth and the Indian summer monsoon: an observational study", *J. Climate*, 12, 3117-3132.

Bhatia R. C., Singh D., Prasad S. and Mukherjee S. K., 2002, "Current status of operationally derived INSAT AMVs quality: details of recent improvements and utilization of derived products", *Proc. Sixth Int. Winds Workshop,* Madison, WI, WMO, 33–43.

Borde R., Hautecoeur O. and Carranza, M., 2016, "EUMETSAT Global AVHRR Wind Product", *J. Atmospheric Oceanic Technol.*, 33, 429-438.

Chelton D. B. and Wentz F. J., 2005, "Global microwave satellite observations of sea surface and sea surface temperature for numerical weather prediction and climate research", *Bull. Amer. Meteor. Soc.*, 86, 1097-1115.

Dash S. K., Singh O. P., Shekhar M. S. and Vernekar A. D., 2003, "Influence of Eurasian snow depth anomaly on the Indian summer monsoon circulation", *Mausam,* 54, 427-442.

Deb S. K., Kishtawal C. M. and. Pal P. K, 2010,"Impact of Kalpana-1-derived water vapor winds on Indian Ocean tropical cyclone forecasts", *Mon. Wea. Rev.*, 138, 987-1003.

Ferraro R. R., 1997, "Special sensor microwave imager derived global rainfall estimates for climatological applications", *J. Geophys. Res: Atmospheres,* 102(D14), 16715-16735.

Fujita T., 1969, "Present status of cloud velocity computation from the ATS-I and ATS-III satellites", *Space Research*, IX, North Holland Pub. Co., Amsterdam, 557-570.

Gadgil S. and Joseph P. V., 2003, "On breaks of the Indian monsoon", *Proc. Indian Acad. Sci. (Earth Planet. Sci*, 112, 529-558.

Gairola R. M., Varma A. K. and Agarwal V. K., 2003, "Rainfall estimation using spaceborne microwave radar and radiometric measurements", *Mausam*, 54, 89-106.

Garg V., 2014, "Outgoing long wave radiation (OLR) a proxy of convection", *International J. Eng. Res. Management Technology*,1, 229-236.

George G. and co-authors, 2013, "Validation of Potential Fishing Zone (PFZ) forecasts from Andaman and Nicobar Islands", *Fishery Technology*, 50, 1-5.

Giri R. K., Yadav R., Ranalkar M., and Singh V., 2022, "Validation of INSAT-3DR sounder retrieved temperature profile with GPS radiosonde and AIRS observations,. *Adv. Space Res.*, 69, 1100-1115.

Gopikrishnan G. S., Kuttippurath J., Thapliyal P. K and Shukla M. V., 2023, Validation of INSAT-3D and INSAT-3DR temperature profile retrievals using ground-based, satellite, and reanalysis data"., *J. Geophys. Res.:* Atmospheres, 128(22), e2023JD038912.

Harries J. E. and Crommelynck D., 1999, "The geostationary earth radiation budget experiment on MSG-1 and its potential application", *Adv. Space Res.,* 24, 915–919.

Harries J. E. et al, 2005, "The geostationary earth radiation budget project", *Bull. Amer. Meteor. Soc.*, 86, 945-960.

Hasler A. F. et al, 1998, "High-resolution wind fields within the inner core and eye of a mature tropical cyclone from GOES 1-min images", *Bull. Amer. Meteor. Soc.,* 79, 2483–2496.

Hawkins G. J. et al, 2008, "High-performance infrared narrow-bandpass filters for the Indian National Satellite System meteorological instrument (INSAT-3D)", *Appl. Optics*, 47, 2346-2356.

Holmlund K., Velden C., and Rohn M., 2001, "Enhanced automated quality control applied to high-density satellite-derived winds", *Mon. Wea. Rev.*, 129, 517-529.

Holmlund K, 1998, "The utilization of statistical properties of satellite-derived atmospheric motion vectors to derive quality indicators", *Wea. Forecasting*, 13, 1093-1104.

IMD, 2014, *INSAT-3D Product catalogue*, 97 pp.

Janowiak J. E., 1992, "Tropical rainfall: a comparison of satellite-derived rainfall estimates with model precipitation forecasts, climatologies and observations", *Mon. Wea. Rev.*, 120, 448-462.

Jin M., 2004, "Analysis of land skin temperature using AVHRR observations", *Bull. Amer. Meteor. Soc.*, 85, 587-600.

Joseph P. V. and Sijikumar S., 2004, "Intraseasonal variability of the low-level jet stream of the Asian summer monsoon", *J. Climate*, 17, 1449-1458.

Joseph P. V., Sooraj K. P. and Rajan C. K., 2006, "The summer monsoon onset process over south Asia and an objective method for the date of monsoon onset over Kerala", *Int. J. Climatol.,* 26.

Kelkar R. R. and Khanna P. N., 1986, "Automated extraction of cloud motion vectors from INSAT-1B imagery", *Mausam*, 37, 495-500.

Kelkar R. R. and Rao A. V. R. K., 1990, "Interannual variability of monsoon rainfall as estimated from INSAT-1B data", *Mausam*, 41, 183-188.

Kelkar R. R., Rao A. V. R. K. and Bhatia R. C., 1993, "Recent improvements in cloud motion vector derivation from INSAT", *Second International Winds Workshop*, Tokyo.

Kelkar R. R., Rao A. V. R. K. and Prasad S., 1993, "Diurnal variation of outgoing longwave radiation derived from INSAT-1B data", *Mausam*, 44, 45-52.

Kelkar R. R., 2015, "The Southern Oscillation was discovered in India, *Ocean Digest*, 2, 8-9

Key J. et al, 2003, "Cloud-drift and water vapor winds in the polar regions from MODIS", *IEEE Trans. Geosci. Remote Sens.*, 41, 482-492.

Khanna P. N., Bhatia R. C. and Singh D., 2000, "Recent improvements in the quality of INSAT-derived AMV and their use in NWP". *Proc. of the Fifth Int. Winds Workshop,* Lorne, Australia, WMO, 103-108.

Khanna P. N. and Kelkar R. R., 1993, "Temperature sounding of the atmosphere over the Indian region using satellite data", *Mausam*, 44, 167-174.

Krishnan R., Zhang C. and Sugi M., 2000, "Dynamics of Breaks in the Indian summer monsoon", *J. Atmos. Sci.*, 57, 1354-1372.

Kulkarni A. V., Mathur P., Rathore B. P., Alex S., Thakur N. and Manoj Kumar, 2002, "Effect of global warming on snow ablation pattern in the Himalayas", Current Sci., 83, 120-123.

Le Marshall J., Leslie L. M., Seecamp R. and Rea A., 2003, "High resolution space based wind observations estimation and application", *Mausam*, 54, 39-50.

Li Z. et al, 2013,"Satellite-derived land surface temperature: Current status and perspectives", *Remote Sensing Environment*, 131, 14-37.

Majumder S. et al, 2021, "Fishing zone characterization in the Indian Ocean by machine learning approach", *Soft Computing for Problem Solving Proc Soc Pro 2020,* 2. Springer Singapore, 43-45.

McMullan K. D. and coauthors, 2008, "SMOS: The payload", *Geosci. Remote Sensing, IEEE Transactions*, 46, 594-605.

McMillin L. M., Gelman M. E., Sanyal A. and Sylva M., 1988, "A method for use of satellite retrieval as a transfer standard to determine systematic radiosonde errors", *Mon. Wea. Rev.*, 16, 1091-1101.

Menzel W. P., 2001, "Cloud tracking with satellite imagery: from the pioneering work of Ted Fujita to the present", *Bull. Amer. Meteor. Soc.*, 82, 33-47.

Milesi C. and coauthors, 2010, "Decadal variations in NDVI and food production in India", *Remote Sensing*. 2, 758-776.

Mitra, A. K. et al, 2015, "INSAT-3D vertical profile retrievals at IMDPS, New Delhi: a preliminary evaluation", *Mausam*, 66, 687-694.

Nieman S., Schmetz J., and Menzel P., 1993, "A comparison of several techniques to assign heights to cloud tracers", *J. Appl. Meteor.*, 32, 1559-1568.

Ohring G., Gruber A. and Ellingson R., 1984, "Satellite determination of the relationship between total longwave radiation flux and infrared window radiance", *J. Clim. Appl. Meteor.*, 23, 416-425.

Pattanaik D. R., Jayanthi N. and Mazumdar A. B., 2005a, "Contrasting pre-monsoon features associated with early and late onset of southwest monsoon over Kerala", *Mausam*, 56, 617-626.

Pattanaik D. R., Kalsi S. R. and Hatwar H. R., 2005b, "Evolution of convection anomalies over the Indo-Pacific region in relation to Indian monsoon rainfall", *Mausam*, 56, 811-824.

Pearce A. and Grundlingh M., 2001, "Applications of satellite remote sensing in the southern Indian Ocean", The Indian Ocean – a perspective", Vol. 1, Oxford IBH Pub., 95-158.

Prasad A. K et al, 2004, "Inter-annual variability of vegetation cover and rainfall over India", *Proc. 35th COSPAR Scientific Assembly*, 18-25 July 2004, Paris, 1015.

Prasad O., Prasad S., Prasad K. and Kelkar R. R., 1998, "Some applications of humidity profiles estimated from INSAT infra-red digital cloud imagery data", *Proc. Indian Acad. Sci., (Earth Planet. Sci.),*107, 5-18.

Prasad S., Rajawat A. S., Pradhan Y., Chauhan O. S. and Nayak S. R., 2000, "Retrieval of sea surface velocities using sequential Ocean Colour Monitor (OCM) data", *Current Sci.,* 80.

Ramanathan, V., Cess R. D., Harrison E. F., Minnis P., Barkstrom B. R., Ahmad E. and Hartmann D., 1989, "Cloud-radiative forcing and climate: results from the earth radiation budget experiment", *Science*, 243, 57–62.

Rao A. V. R. K., Kelkar R. R. and Arkin P. A., 1989, "Estimation of precipitation and outgoing longwave radiation from INSAT-1B radiance data", *Mausam*, 40, 123-130.

Rao B. M. and coauthors, 2001, "Large scale soil moisture estimation using microwave radiometer data", *J. Agrometeor.*, 3, 179-187.

Rao, V. K. et al, 2020, "Validating INSAT-3D atmospheric temperature retrievals over India using radiosonde measurements and other satellite observations", *Meteor. Atmos. Phys.*, 132, 583-601.

Sant Prasad, Khanna P. N., Rao A. V. R. K. and Kelkar R. R., 1990, "Satellite-derived monthly average wind fields over the Indian Ocean in April-July 1988", *Mausam*, 41, 445-450.

Shyamala B., 1997, "Satellite studies of monsoon processes", *Proc. First Int. Workshop on Monsoon*, Bali, Indonesia.

Shyamala B and Sudevan S., 1999, "Objective criteria for withdrawal of southwest monsoon based on OLR data", *Mausam*, 50, 406-408.

Smith W. L. et al, 1979, "The TIROS-N operational vertical sounder", *Bull. Amer. Meteor. Soc.*, 60, 1177-1187.

Sobrino, J. A., Li Z.-L., Stoll M. P. and Becker F., 1994, "Improvements in the split-window technique for land surface temperature determination", *IEEE Trans. Geosci. Remote Sens.,* 32**,** 243–253.

Solanki H. U., Dwiwedi R. M. and Nayak S. R., 2001, "Synergistic analysis of SeaWiFS chlorophyll concentration and NOAA-AVHRR SST features for exploring marine living resources", *Int. J. Remote Sensing,* 22, 3877-3882.

Solanki H. U. and coauthors, 2003, "Fishery forecast using OCM chlorophyll concentration and AVHRR SST: validation results off Gujarat coast, India", *Int. J. Remote Sensing,* 24, 3691-3699.

Srivastava A. K. and Rajeevan M., 2004, "Prediction of summer monsoon rainfall over northwest region of India using an OLR index", *Mausam*, 55, 583-590.

Srivastava A. K., Rajeevan M. and Kulkarni R., 2002, "Teleconnection of OLR and SST anomalies over Atlantic Ocean with Indian summer monsoon", *Geophys. Res. Letters*, 29, 125.1-4.

Suresh R., 2002, "On forecasting cyclone movement using TOVS data", *Mausam*, 53, 215-224.

Suresh R., 2002, "On variability of total ozone derived from TOVS data over peninsular Indian sub-continent and adjoining oceanic area", *Mausam*, 53, 503-514.

Suresh R. and Raj Y. E. A., 2001, "Some aspects of Indian northeast monsoon as derived from TOVS data", *Mausam*, 52, 727-732.

Suresh R., Sankaran P. V. and Rengarajan S., 2002, "Atmospheric boundary layer during northeast monsoon over Tamil Nadu and neighbourhood - a study using TOVS data", *Mausam*, 53, 75-86.

Thapliyal P. K., Rao B. M., Pal P. K. and Das H. P., 2003, "Potential of IRS-P4 microwave radiometer data for soil moisture estimation over India", *Mausam,* 54, 277-286

Thapliyal P. K., Pal P. K., Narayanan M. S. and Srinivasan J., 2005, "Development of a time series based methodology for estimation of large area soil Wetness over India using IRS-P4 microwave radiometer data", *J. Appl. Met.*, 44, 127-143.

Velden C., Stettner D. and Daniels J,, 2000, "Wind vector fields derived from GOES rapid scan imagery", Preprints, *10th Conf. on Satellite Meteorology*, Long Beach, CA, Amer. Meteor. Soc., 20–23.

Velden C. et al, 2005, "Recent innovations in deriving tropospheric winds from meteorological satellites", *Bull. Amer. Meteor. Soc.*, 86, 205-223.

Walker G. T., 1923, "Correlation in seasonal variations of weather VIII: A preliminary study of world weather", *Memoirs India Meteor. Dept.,* 24, 75-132.

Wentz F. J., Gentemann C., Smith D., and Chelton D., 2000, "Satellite measurements of sea surface temperature through clouds", *Science*, 288, 847–850.

Wielicki, B. A. et al, 2002, "Evidence for large decadal variability in the tropical mean radiative energy budget", *Science,* 295,841–844.

Yadav B. R. and Kelkar R. R., 1989, "Lower level wind flow over the Indian Ocean during the onset of monsoon-1987", *Mausam*, 40, 97-104.

Yadav R., Giri R.K. and Bhan S.C., 2022, "High-resolution outgoing long wave radiation data (2014–2020) of INSAT-3D Imager and its comparison with Clouds and Earth's Radiant Energy System (CERES) data", *Advances in Space Res.,* 70, 976-991.

Chapter 4

Measurement of Parameters by Active Sensors

Historically, meteorological observations have always been made with sensors that respond to the environment in a passive manner. A revolutionary change in this traditional method of observation came about after the Second World War in the mid-twentieth century, when military radars which had been deployed for spotting enemy targets and airplanes were offered to meteorologists as disposal items. Radars have since come into use in a wide variety of applications in everyday life, and the weather radar has become the operational forecaster's primary tool for viewing rapidly evolving local weather and storms out at sea.

A radar (an abbreviation of RAdio Detection and Ranging) sends out pulses of electromagnetic radiation and integrates the returned energy over a large volume. The transmitted pulse is reflected back to the radar by an object or target that may be in its path. The time taken by the pulse to return to the radar is a measure of the range or distance of the target from the radar. The characteristics of the returned pulse or echo can be different from those of the transmitted pulse and are an indicator of the size and nature of the target. In a Doppler radar the speed of movement of the target can also be estimated by measuring the Doppler frequency shift.

Weather radar echoes from clouds are very weak, but rain and hail are very easily detected by most radars and quantitative estimation of rain is possible. They are extremely useful is observations of local severe storms, mesoscale convective complexes, tornadoes and tropical cyclones near the coast. Clear air returns depend on the particulate matter in the atmosphere. As atmospheric refractive index gradients are dependent on the vertical profiles of temperature and humidity, it is possible to estimate wind speeds, wind shear and turbulence.

Radars enable the computation of the precipitation through what are known as Z-R relationships. Their general from is $Z = A\ R^b$, R being the rainfall and Z the radar reflectivity factor or the power returned. The constant b has a numerical value in the range 1-2 and constant A in the range 100-300. The best known relationship is $Z=200\ R^{1.6}$ (Raghavan 2003).The numerous Z-R relationships are the result of the use of different assumptions with regard to micro-physical characteristics like drop size distribution, presence of hail, coalescence of drops and evaporation of drops below the cloud base. The relationships are also different because different statistical methods have been used for averaging gauge data over an area. The relationships show a general bias towards larger rain drop sizes and they may be affected by radar errors like beam blockage in hilly terrain and attenuation due to intervening rainfall.

Rain gauges on the other hand measure the rainfall amount accumulated over a period of time. For rain rate and intensity measurement, we need to take frequent rain gauge readings or have automatic self-recording gauges. Because of its high variability, rainfall averaging over large areas requires a very dense rain gauge network. Radars give areal rainfall and frequent measurement is possible making them particularly useful for flood forecasting. However, because of their long and widespread use, rain gauges continue to remain as the standard for measuring surface rainfall and are used as ground truth in radar calibration.

The growth of both radar technology and radar meteorology over the last few decades has indeed been phenomenal, and we now have various types of atmospheric radars operating in different frequencies and serving different areas of applications, in research as well as operations (Raghavan 2003). The X-, C-, and S-band microwave radars are used for purposes of storm detection and measurement of rain drop size and rain rate. Sodars, which use acoustic or sound waves, have applications in studies of atmospheric turbulence in the boundary layer. Lidars operate with optical or light waves and are coming into increasing use for measurement of atmospheric aerosols. Millimetre wave radars provide measurements of cloud physical properties and vertical structure.

Ultra and Very High Frequency radars like the MST Radar have been established at many locations around the globe for obtaining vertical profiles of winds and other atmospheric parameters. The abbreviation MST is for the mesosphere, stratosphere and troposphere, indicating that the radar beam is strong enough to penetrate these atmospheric layers. India has an MST radar located at Gadanki, near Tirupati. It operates at a frequency of 53 MHz corresponding to a wavelength of 5.66 m. It uses the Doppler beam swinging technique to get the three-dimensional wind field. The vertical resolution is as high as 150 m and the time resolution is between 1 and 2 min (Rao et al 1995). The MST radar gives direct measurements of vertical velocities in the atmosphere. Asnani et al (2000) have analyzed MST radar profiles which show the vertical velocities changing sign every few kilometres of height above ground, leading to the formation of thin layered cirrus clouds particularly over the tropopause.

While the possibility of launching radars on satellites was considered since long, there were great difficulties in realizing it and meteorological remote sensing worked essentially in a passive mode. The fast motion of satellites in low elevation orbits, the extremely long range at geostationary height, and considerations of radar power, weight and antenna size, posed daunting challenges in the way of putting a weather radar in space. The concept of a space-borne precipitation radar mooted way back in 1984 was realized thirteen years later with the launch of the TRMM satellite in 1997. With this development, a new era of active meteorological remote sensing has begun and active sensor technology is now proliferating. Many satellites are already making new types of measurements demanded by meteorological, oceanographic and climate change related applications that passive sensors are unable to cater to, and more satellites are planned to be launched in the near future. These developments are covered in the subsequent sections of this chapter.

4.1 Precipitation

Prior to the satellite era, we did not have a comprehensive knowledge of the global rainfall distribution as it was not possible to establish rain gauges over the wide oceans. Rain measurement on ships was also not easy. Maintaining rain gauges over some land areas like mountains and deserts was equally difficult because of logistic problems. There were very few precipitation radars around the globe than what were required, because of their high cost.

The estimation of large-scale precipitation from geostationary IR radiances was a major breakthrough in improving our understanding of global rainfall patterns, particularly over the oceans, but this method is essentially statistical and does not work on smaller space and time scales and has additional errors over land. Statistical techniques were also developed for deriving rainfall from measurements made by passive microwave sensors, but it was found that different algorithms produced very diverse rain estimates.

The classical view regarding the diurnal variation of precipitation has been that over continental stations, rainfall is maximum in the afternoon and has a convective nature, and that over coastal and island stations, it is maximum at night or in the early morning. However, there are many exceptions and terrain irregularities cause intense mesoscale diurnal circulations which alter the pattern (Asnani 2005). Actual rainfall observations over the oceans are too meagre to arrive at any firm conclusions and only satellite studies can throw light on the diurnal precipitation variation over the oceans (Murakami 1983).

In summary, although meteorological satellites have given us a great deal of new information about global rainfall, many aspects such as rainfall amounts over the tropical oceans, the diurnal cycle of precipitation, the rain rate and its vertical profile, have continued to remain largely uncertain and unvalidated. At the same time, since a major contribution to the energy that drives the atmospheric circulation comes from the latent heat released by tropical precipitation, climate modellers have demanded a far greater accuracy in the precipitation measurements that go as inputs into their models.

The idea of measuring rainfall from a satellite having a combination of passive sensors and a microwave radar instrument was mooted in the early 1980s and it took the shape of a proposal to launch a Tropical Rainfall Measuring Mission (TRMM). The TRMM project went through a long process of evolution and eventually it became a joint mission of NASA and the Japan Aerospace Exploration Agency (JAXA). TRMM became the first satellite to have an on-board radar to measure precipitation quantitatively from space.

Several orbit and altitude options were considered and it was decided to place the satellite at an altitude of 350 km in an inclined orbit extending between 35°N and 35°S. The inclined orbit precessed in such a manner that the satellite would fly over a given location at a different time every day with a repeat cycle of about two days. This orbit would allow the measurement of the large diurnal variation of tropical rainfall. The TRMM satellite was launched in November 1997.

4.1.1 TRMM and GPM – Radar on a Satellite

For placing a radar on board a satellite, certain basic problems need to be overcome. For a radar on a geostationary satellite, the range becomes enormous. On the other hand, polar orbiting satellites move at high speed relative to the earth and the dwell time is insufficient. Therefore, the beam width of the radar has to be narrow enough to be fully filled with precipitation and the swath has to be wide enough to eliminate gaps in coverage (Varma et al 2004). A space-borne radar must also be able to discriminate the weak precipitation echo over the strong return from the earth's surface. It is hence desirable to use wavelengths shorter than 3-10 cm, which are used in conventional weather radars, as at shorter wavelengths the return echo is stronger.

The basic payload of the TRMM satellite was the Precipitation Radar (PR). It also carried a 9-channel passive TRMM Microwave Imager (TMI) which was similar to the SSM/I, and a Visible Infrared Scanner (VIRS) which made measurements in the VIS and IR channels like the AVHRR. Besides these three primary payloads, TRMM had two other instruments, the Clouds and the Earth's Radiant Energy System (CERES), and the Lightning Imaging Sensor (LIS). PR, TMI and VIRS were designed to obtain rainfall and other relevant information such as rain type, height of the bright band, cloud type and cloud top height individually and complement each other's functions. Japan developed the precipitation radar and launched the satellite, while the U.S. provided the other sensors and systems.

A detailed description of the three primary TRMM payloads, PR, TMI and VIRS, has been given by Kummerow et al (1998). The PR was a 128-element phased array system operating at two frequencies, 13.796 and 13.802 GHz. The PR had a horizontal resolution of 4.3 km at nadir and a swath width of 215 km. One of its most important features was its ability to provide vertical profiles of rain and snow from the surface up to a height of 20 km at a vertical resolution of 250 m. The PR could perceive rain through clouds and detect even light rain rates of the order of 0.5 mm/hr. At intense rain rates, where the attenuation effects can be strong, new methods of data processing were developed to help correct for this effect. The power requirement of the PR was incredibly low, just 224 watts.

One of the earliest attempts to use TRMM rainfall information to estimate the four-dimensional latent heating structure over the global tropics was made by Tao et al (2001). Later on, Adler et al (2003) analyzed three years of TRMM monthly estimates of tropical surface rainfall to document and understand the differences among the TRMM-based estimates and compare them with pre-TRMM estimates and current operational analyses. Variation among the TRMM estimates was shown to be considerably smaller than with pre-TRMM collection of passive microwave-based products. TRMM estimates were shown to have a range of about 20% for the tropical ocean as a whole and 30% over heavily raining ocean areas of the ITCZ. In mid-latitude ocean areas the differences were smaller. Over land there was a distinct difference between the tropics and middle latitudes.

The diurnal variation of precipitation has been studied by Nesbitt et al (2003) using three years of TRMM data. They found that over the oceans, the diurnal cycle of rainfall has a small amplitude, with the maximum contribution to rainfall coming from mesoscale convective systems (MCSs) in the early morning. Land areas have a much larger rainfall cycle than over the ocean, with a marked minimum in the mid-morning hours and a maximum in the afternoon, slowly decreasing through midnight. However, all land regions have MCS rainfall peaks that occur in the late evening through midnight due to their longer life cycle. The diurnal cycle of MCS rainfall over land and convective intensity varies significantly from one region to another.

Hirose et al (2005) investigated the spatial and diurnal variation of rainfall over Asia using TRMM PR data for 1998-2003. Small precipitation systems occurred most frequently around early afternoon over land. The south-facing slopes of the Himalayas, especially the upper portion of the Brahmaputra valley, are the most obvious region of the daytime genesis of the convective systems over the Asian landmass. Over the Tibetan Plateau the occurrence of the small systems was larger than over inland India and the foothills. Large systems developed mostly in the evening over nearly flat landmasses. Widespread systems with intense rain pixels developed over the foothills of the Himalayas in the late night to early morning period, which was distinct from the daytime convection.

The TRMM Precipitation Radar was the first space-borne instrument to provide three-dimensional maps of tropical storm structure. The measurements yielded invaluable information on intensity, distribution and type of the rain and the height at which the snow melts into rain. The TRMM Microwave Imager (TMI) measured the amount of energy scattered by the ice particles. The more ice there is, the more energy is scattered, meaning smaller amounts of energy are received by TRMM and lowest values of returned energy (200° K) represent the most intense clouds. By following a sequence of TRMM images, changes in the structure of the eyewall and rain bands with time can be analyzed. Many times the evolution of a double eyewall can be clearly seen. These signals help in estimating whether a storm will intensify or weaken.

Cecil et al (2002) documented 261 overpasses of 45 hurricanes by the TRMM satellite during 1997-98, with the observed radar reflectivity values, passive microwave ice scattering magnitudes, and total lightning. Properties of the three regions of the hurricane, eyewall, inner rain band and outer rain band, were treated separately and compared to other tropical oceanic and tropical continental precipitation systems. Reflectivity profiles and ice scattering signatures are found to be fairly similar for both hurricane and non-hurricane tropical oceanic precipitation systems, although the hurricane inner rain band region yields the weakest of these convective signatures. When normalized by the area experiencing significant convection, the outer rainband region produces more lightning than the rest of the hurricane or non-hurricane tropical oceanic systems. As a whole, the tropical oceanic precipitation systems, both hurricane and non-hurricane, are dominated by stratiform rain and relatively weak convection.

The TRMM mission came to an end in 2015 and was replaced by the Global Precipitation Mission which is further described in Section 5.3.

The Dual-frequency Precipitation Radar (DPR) consists of a Ku-band precipitation radar (KuPR) and a Ka-band precipitation radar (KaPR). The KuPR operating at 13.6 GHz is an advanced version of the PR flown on TRMM. The KaPR is also basically similar in design to the TRMM PR but it has a frequency of 35.55 GHz in the Ka-band. The phased array antennas of both KuPR and KaPR are co-aligned on the GPM spacecraft bus in such a manner that both of them have identical 5 km size footprints on the ground.

Since the precipitation radars operate at two widely separated frequencies, the attenuation and reflectivity of the transmitted pulses would be different.

This information permits a better differentiation of drop size distributions and rain rates. The GPM DPR is more sensitive than the TRMM PR especially in the measurement of light rainfall and snowfall in the high latitude regions. The variable pulse repetition frequency technique is adopted in order to realize a sensitivity of 0.2 mm/hr in the rain rate measurement.

The GPM Microwave Imager (GMI) instrument is a multi-channel, conical-scanning, microwave radiometer that serves as a precipitation standard as well as a radiometric standard for the other GPM constellation members. The GMI has 13 microwave channels ranging in frequency from 10 to 183 GHz. Besides carrying the same five channels as the TRMM Microwave Imager (TMI), the GMI also carries four high frequency, millimetre-wave channels in the 166 to 183 GHz range. The DPR and GMI swaths have the necessary overlap required for radiometer-based precipitation retrievals.

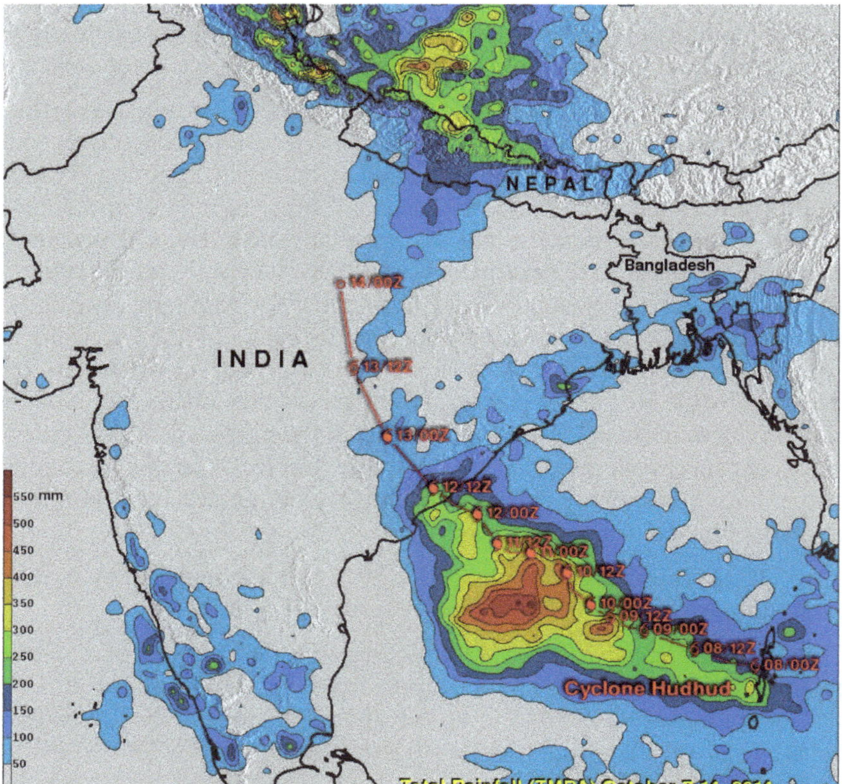

Figure 4.1.1.1 TRMM rainfall accumulation along the track of cyclone Hudhud over Bay of Bengal 7-14 October 2014.
(Source: https://gpm.nasa.gov/mission-updates)

Figure 4.1.1.1 shows the near real time TRMM-based Multi-satellite Precipitation Analysis (TMPA) data along the track of tropical cyclone Hudhud over the Bay of Bengal for the period 7-14 October 2014. Hudhud had reached the equivalent of a category 4 hurricane on the U.S. Saffir-Simpson scale but weakened slightly before making landfall near Visakhapatnam on the east coast of India. Highest TMPA rainfall totals of the order of 550 mm or more (dark red) are seen over the open ocean. Rainfall totals are 200 to 250 mm (green) over the landfall region and 50 to 100 mm (blue) further inland.

4.2 Cloud Properties

In the early decades of its evolution, satellite meteorology had focused on the observation of global cloud cover and interpretation of cloud patterns as indicators of the development and movement of weather systems. Only by the end of the twentieth century clouds came to be considered as not just passive tracers or end products of atmospheric processes, but as playing an independent and dominant role in them. Changes in the global cloud distribution can possibly affect the future climate, perhaps even more than what greenhouse gases and anthropogenic aerosols may do (Wielicki et al 1995, Glecker et al 1995).

Clouds have a crucial role in the hydrological cycle, acting as the link between the condensation of atmospheric water vapour and occurrence of precipitation. They develop in regions of high relative humidity and in turn affect the moisture distribution. Clouds are also a major component of the earth's radiation budget because as clouds form, condensation releases latent heat. They reflect solar radiation back to space, and absorb and emit terrestrial thermal radiation. Any long term change in cloud distribution will therefore result in modifying the distribution of heating within the atmosphere. It is further possible that changes in climate may in turn give rise to changes in clouds and these feedbacks could be either positive or negative.

Climate models need to have a resolution of cloud layers and their radiative properties which is high enough for an accurate assessment of the poleward transport of heat (Stocker et al 2001). The cloud cover imagery and vertically integrated products provided by the conventional type of meteorological satellites are insufficient for this purpose. Entirely different satellite instruments are required for making measurements that would reveal the vertical structure of clouds.

While the three-dimensional geometrical properties of clouds like fractional cloud cover and cloud boundaries directly enter the process of solving the radiative transfer equations, microphysical properties of the hydrometeors they contain, like ice or water phase, particle size, shape, non-sphericity and orientation, and fall velocity of cloud particles, are no less important. The cloud feedbacks produced by atmospheric general circulation models differ both in magnitude and sign depending on the manner in which the cloud optical properties are treated (Yao et al 1999). The following sections describe the recent advances in space-based measurements of cloud properties by dedicated satellite missions that complement the general imaging capabilities of meteorological satellites.

4.2.1 POLDER – Lidar on a Satellite

The visible solar radiation coming from the sun is not polarized, but certain physical processes on earth result in generating polarized light. Satellite imageries in the VIS channel show bright areas called sunglint over calm oceans at certain times. They are less apparent when the sea surface is rough (Section 2.1). However, in addition to reflection, polarization of sunlight also takes place, which can be measured by a satellite.

The measurement of polarization which is not done by the operational meteorological satellites, helps to clearly identify the cloud phase, because liquid droplets have a highly specific signature in a preferred viewing direction. Polarization data, when combined with directionality, helps to determine the physical and optical properties of aerosols and clouds and their influence on the earth's radiation budget (Bellouin et al 2003).

An instrument for measuring the POLarization and Directionality of the Earth's Reflectances (POLDER) was conceived and developed by CNES in France. POLDER-1 was launched on the Japanese Advanced Earth Observation Satellite (ADEOS-I) in August 1996 and POLDER-2 was launched on the ADEOS-II satellite in December 2002. However, the missions had lifetimes of less than a year as both the satellites encountered serious technical glitches. Subsequently, another instrument, POLDER-3, was launched in December 2004 on the French PARASOL microsatellite. PARASOL is an acronym for Polarization and Anisotropy of Reflectances for Atmospheric Sciences coupled with Observations from a Lidar. The PARASOL/POLDER mission far exceeded its design life and operated for nine years up to the end of 2013.

The POLDER instrument had a camera comprising a two-dimensional CCD detector array, wide field of view optics and a rotating wheel carrying spectral and polarized filters. POLDER had a total of 9 spectral bands, out of which three bands, 443, 670 and 865 nm, had polarized filters (Buriez et al 1997, Deuze et al 2000). Both the ADEOS satellites were in a sun-synchronous orbit with a 10:30 am descending node, flying at an altitude of 800 km and having a repeat cycle of 4 days. PARASOL had an equator crossing time of 1:30 pm and it was a part of the A-Train for some years (Section 5.4).

4.2.2 CloudSat

The basic premise of remote sensing is that by viewing the same object or surface in different bands or regions of the electromagnetic spectrum, different properties of the object come to light. A VIS image of a cloud indicates the reflection of solar radiation from its top, whereas an IR image indicates the temperature and so the height of the cloud top. Likewise if a cloud could be scanned simultaneously by a radar and a lidar, different regions of the cloud would respond with different sensitivity to the centimetre or millimetre wavelength of the radar and the nanometre wavelength of the lidar. The relatively wider beam of the radar would be more sensitive to large particles, while finer particles would be better observed by the narrow beam of the lidar. The radar and lidar can thus be used together so as to complement each other's capabilities and provide a more complete description of the cloud top, base and depth with a higher precision than they can do individually.

The drop size distribution of a cloud depends on several factors such as the updraft velocity, concentration of cloud condensation nuclei and the height above the cloud base. The optical depth and albedo of a cloud are a function of the water content and effective radius. Formation of precipitation by the warm cloud process requires that some drops grow large enough for efficient coalescence. Satellite measurement of cloud properties became a reality with the successful launch of two satellites, CloudSat and CALIPSO, in April 2006. Although initially expected to last only for a couple of years, both satellites continued to work until 2023.

CloudSat carried a mm-wave radar and CALIPSO had on board a backscattering lidar. The pair of satellites formed a unique combination of radar and lidar in space. Stephens et al (2002) have given an excellent overview of the CloudSat mission, onboard instruments, data processing and validation plans.

The science objectives of CloudSat were to observe the vertical structure of clouds, measure the profiles of cloud liquid water, ice water content and optical properties. They helped to determine how clouds affect both the local and large-scale environment, and to validate the results of cloud process models and cloud parameterizations in global scale models.

Ground-based weather radars generally operate at wavelengths of 2 to 15 cm in the X, C and S bands to enable the detection of raindrop-sized particles. Radars operating at 3 or 8 mm (94 or 35 GHz) can detect the much smaller particles of liquid water and ice that constitute the large cloud masses. The CloudSat Cloud Profiling Radar (CPR) which was developed jointly by NASA/Jet Propulsion Laboratory and the Canadian Space Agency was, however, the first mm-wavelength cloud radar to fly in space. The CPR was a 94 GHz nadir-looking radar which measured the power backscattered by clouds as a function of distance from the radar. The orbit of CloudSat was identical to the sun-synchronous, 705 km altitude, polar orbit of the Aqua satellite.

Rajeevan et al (2013) made a study of vertical cloud structure of the Indian summer monsoon using four years (2007-2010) of CloudSat CPR data on column integrated and vertical profiles of cloud water and ice water content. They observed deep convective clouds extending up to 14 km and large values of cloud water and ice content over north Bay of Bengal, north Arabian Sea and adjoining northwest India, along the west coast of India and south equatorial Indian Ocean. Over the Bay of Bengal, cloud liquid water content and ice water content was highly enhanced during active monsoon spells.

Figure 4.2.2.1 shows an Aqua MODIS image of Category 4 Typhoon Dolphin over the Pacific Ocean on 16 May 2015 with a CloudSat CPR reflectivity overlay. CloudSat had traversed the system directly over the eye, and one can clearly see the outward sloping eyewall, intense convection and rainfall in the surrounding eyewall, and the storm's overall cloud structure.

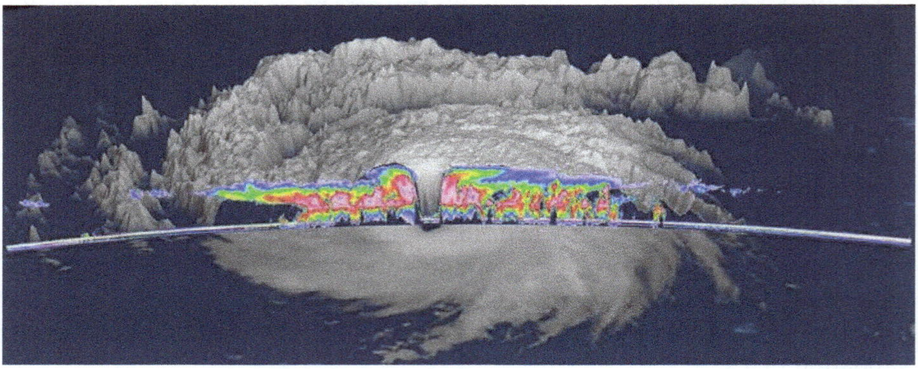

Figure 4.2.2.1 Aqua MODIS and CloudSat image 16 May 2015 of Typhoon Dolphin.
(Source: https://www.jpl.nasa.gov/images/)

4.2.3 CALIPSO

The Cloud-Aerosol Lidar and Infrared Pathfinder Satellite Observation (CALIPSO) mission was launched in April 2006 simultaneously with CloudSat as mentioned in the preceding section. CALIPSO was a joint U.S.-French (NASA-CNES) enterprise and the satellite far exceeded its expected life of three years. It has provided a new insight into the role that clouds and atmospheric aerosols play in regulating the earth's weather, climate and air quality.

The CALIPSO payload consists of three co-aligned nadir-viewing instruments. The primary payload was the Cloud-Aerosol Lidar with Orthogonal Polarization (CALIOP), which was a dual-wavelength polarization-sensitive Nd YAG lidar that provided high resolution vertical profiles of aerosols and clouds. CALIOP measured the 1064 nm backscatter intensity and had two more receiver channels for measuring orthogonally polarized components of the 532 nm backscattered signal. The receiver telescope was 1 m in diameter. Measurement of the linear depolarization of the backscattered return allowed discrimination of the cloud phase and detection of non-spherical aerosols.

The second payload on CALIPSO was the Imaging Infrared Radiometer (IIR) that provided calibrated IR radiances at wavelengths of 8.65, 10.6 and 12.0 μ. IIR was a nadir-viewing, non-scanning imager having a 64 × 64 km swath with a pixel size of 1 km. The CALIOP beam was nominally aligned with the centre of the IIR image. The three wavelengths were selected to optimize joint CALIOP/IIR retrievals of cirrus cloud emissivity and particle size.

CALIPSO also had a high-resolution Wide Field Camera (WFC) that acquired high spatial resolution imagery. It was a fixed, nadir-viewing imager with a single spectral channel covering the 620-670 nm region, selected to match Band 1 of the MODIS instrument on Aqua.

The CALIOP lidar operated on the same principle as a radar except that the laser transmits pulses of green light and IR energy to the target instead of microwave energy. The transmitted light interacts with and is changed by the target. Some of this light is reflected or scattered back to the instrument. The change in the properties of the light enables some property of the target to be determined. The time for the light to travel out to the target and back to the lidar is used to determine the range to the target.

CALIOP had a basic vertical resolution of 30-60 m and a horizontal resolution of 333 m. However, the atmosphere becomes more spatially uniform with increasing altitude, and signals from higher levels in the atmosphere tend to be weaker. Therefore, an altitude-dependent on-board averaging scheme was developed which provides full resolution in the lower troposphere, where the spatial variability of cloud and aerosol is greatest, and lesser resolution in the upper atmosphere.

Detailed overviews of the CALIPSO mission and CALIOP data processing algorithms have been given by Winker et al (2003, 2004, 2009). The fundamental science objective of the CALIPSO mission was to provide vertically resolved information on aerosol distribution, extinction coefficient, hydration state, discrimination of large and small particles, and cloud masking, in order to enable climate models to estimate direct as well as indirect aerosol forcing and reduce uncertainty. Another science objective of CALIPSO was to provide comprehensive observations of cloud height, thickness and multi-layering, for more accurate computation of surface and atmospheric radiative fluxes in global climate models.

Hazra et al (2015) carried out an evaluation of cloud properties in the NCEP CFS v2 model and its linkage with the Indian summer monsoon. Cloud fraction, which varies greatly among general circulation models, plays a crucial role in simulation of Indian summer monsoon rainfall. CFS v2 was evaluated in terms of its simulation of cloud fraction, cloud condensate, outgoing longwave radiation and tropospheric temperature. Biases in these simulated quantities were computed using CALIOP data.

Srivastava et al (2016) studied vertical profiles of aerosol backscatter coefficient and depolarization ratio, obtained from the CALIPSO satellite, in the upper troposphere and lower stratosphere region over North India covering the Indo-Gangetic Plains for a 1-year period from December 2011 to November 2012. An enhanced aerosol layer was observed between 15 and 18 km altitude, in the vicinity of the tropopause, with a broad layer depth of about 2 km. The aerosol layer showed strong seasonal, monthly as well as day and night time variability, with a peak value of backscatter coefficient during monsoon season. Das et al (2013) also carried out an investigation of the radiative effects of the optically thick dust layer over south India.

4.3 Sea Surface Winds – Scatterometer

Winds in the free atmosphere can be estimated by viewing the motion of individual clouds or moisture tracers in successive images of geostationary satellites. However, this technique cannot be adopted for deriving winds in the planetary boundary layer in which other turbulent forcings are active. In the case of winds over the sea surface, alternative means such as a scatterometer have to be used.

Scatterometers have significantly enhanced the ability to detect and track severe storms and cyclones by providing direct measurements of winds at the sea surface. They have vital applications in various fields such as studies of air-sea interaction, ocean circulation, deforestation or changes in polar sea ice masses. Scatterometer measurements of ocean surface wind speed and direction have also been proved useful in improving the forecast accuracy of numerical weather prediction models by providing the initial fields particularly in the region of tropical storms.

A scatterometer is a radar-like active microwave sensor on an aircraft or a satellite which can measure the reflection or scattering effect produced while scanning the surface of the earth. The measurements can be made equally well in cloudy and clear conditions. A rough ocean surface returns a stronger signal because the waves reflect more of the radar energy back toward the scatterometer antenna. A smooth ocean surface returns a weaker signal because lesser energy is reflected. Thus by measuring the backscatter from the ocean, the scatterometer can estimate the surface wind velocity.

At a given angle to the flight path of the satellite, the amount of backscatter depends on the size of the surface ripples of the ocean, i.e. surface roughness, and their orientation with respect to the direction of propagation of the pulse of radiation transmitted by the scatterometer. The first factor is dependent on the surface wind stress while the second is related to the wind direction. Hence both wind speed and direction could be derived.

Sarkar (2003) has given an exhaustive review of space-based techniques for measurement of sea surface winds, including the use of passive microwave sensors like SSM/I for derivation of sea surface wind magnitude. Another review of the use of scatterometer data over the oceans and its emerging applications over land and ice surfaces, has been made by Liu (2002).

4.3.1 Oceansat-3

The Indian Space Research Organisation (ISRO) has a continuing programme of launching satellites for oceanographic and related atmospheric applications. Oceansat-1 or IRS-P4 was the first Indian satellite mission dedicated to ocean applications and it was launched in May 1999. The satellite carried two payloads, an Ocean Colour Monitor (OCM) and a Multi-frequency Scanning Microwave Radiometer (MSMR). It lasted for 11 years until August 2010.

The second Indian satellite in this series, Oceansat-2, was launched into a near-polar sun-synchronous orbit of 730 km altitude in September 2009. Oceansat-2 carried another OCM for the sake of continuity and a wind scatterometer called OSCAT. This was a pencil-beam scatterometer operating in the Ku-band at 13.73 GHz. It had a conically scanning antenna to transmit microwave pulses and receive the backscattered power from the ocean surface from which wind vector information could be derived. OSCAT operated with two beams, the inner beam in horizontal polarization and the outer beam in vertical polarization. The resolution was 45 km. With a swath of 1800 km OSCAT could cover 90% of the earth's ocean surface in 24 h, which was a substantial improvement over the coverage of other side-looking scatterometers. Unfortunately, OSCAT suffered a fatal anomaly in February 2014 and became unusable thereafter.

Oceansat-3 was launched by ISRO in November 2022 to provide continuity to the earlier two satellites in the Oceansat series with enhanced application potential. It flies in a sun-synchronous polar orbit at 730 km altitude and a repeat cycle od two days. Oceansat-3 has three payloads:

(a) 13-band advanced Ocean Colour Monitor (OCM-3)

(b) Ku-band Scatterometer (SCAT-3)

(c) 2-channel Sea Surface Temperature Monitor (SSTM)

OCM-3 is a 13-band multi-spectral radiometer operating in the VIS and NIR spectral range with 10-20 nm resolution. This provides an instantaneous geometric field of view of 368 m and a swath of 1500 km. It can be tilted up to 20° along track. OCM-3 can be operated in two modes: Local Area Coverage (LAC) mode at 366 m ground resolution and Global Area Coverage (GAC) mode at 1.1 km resolution.

SCAT-3 is an active Ku-band, microwave device to determine ocean surface level wind vectors through estimation of radar backscatter. It has a swath of 1400-1800 km. It can derive surface wind vectors at a resolution of 12.5 or 25 km.

SSTM operates in two bands 10.75-11.25 μ and 11.75-12.25 μ and is used for SST retrievals.

Figure 4.3.1.1 shows the winds measured on 16 August 2025 by the SCAT-3 scatterometer when Oceansat-3 was passing over the North Atlantic Ocean and hurricane Erin was fast intensifying into a Category 4 storm.

Figure 4.3.1.1 Hurricane Erin winds measured by Oceansat-3 scatterometer on 21 August 2025 over North Atlantic Ocean.
(Source: CIMMS web page https://tropic.ssec.wisc.edu/)

Currently two European operational polar meteorological satellites, MetOp-B and MetOp-C, carry the Advanced Scatterometer (ASCAT). This is a C-band scatterometer operating at 5.255 GHz. Gelsthorpe et al (2000) and Figa-Saldana et al (2002) have described the ASCAT system and its capabilities, including the instrument, calibration, ground processing, and products, and reviewed the science associated with scatterometer measurements.

4.4 Sea Surface Height – Altimeter

Altimeters have been in use since long by aircraft pilots to find out the height at which they are flying, above the ground or ocean. Geosat was the first satellite to carry successfully a radar altimeter in 1985 and it worked for more than four years. Although the working principle is the same, the aim of a satellite-based altimeter is not to fix the height of the satellite but to determine where the sea level is. Satellite altimetry opened up an entirely new range of ocean applications, particularly for monitoring of the sea level, and to better understand the ocean circulation.

4.4.1 TOPEX/Poseidon

TOPEX/Poseidon was launched in August 1992 as a joint NASA-CNES programme. The satellite was placed at an altitude of 1336 km, in a circular, non-sun-synchronous, 66° inclination orbit covering 95 % of the world's ice-free oceans between 66°N and 66°S latitudes with a 10-day repeat cycle. The satellite carried two radar altimeters, NASA TOPEX operating at two frequencies 13.6 and 5.3 GHz in the Ku- and C-band respectively, and CNES Poseidon operating at 13.65 GHz in the Ku-band. It also carried a 3-channel NASA microwave radiometer to measure total water vapour along the altimeter path and use it to correct for pulse delay.

The NASA Global Positioning System Payload (GPSP) receiver provided precise orbit and ephemeris data. An essential accompaniment was the CNES DORIS system (Doppler Orbitography and Radio-positioning Integrated by Satellite) which received signals from ground beacons for precise orbit determination, satellite tracking and ionospheric correction data for the CNES altimeter. The NASA Laser Retroreflector Array (LRA) payload served as a target for laser ranging stations on ground stations to track the satellite, calibrate the other satellite location systems and verify altimeter measurements.

The range measurements after correction for atmospheric and instrumental effects, were accurate to 3-4 cm. The range measurements when subtracted from estimates of the satellite orbital height, resulted in ocean height measurements of 4-5 cm accuracy relative to the centre of the earth. This accuracy figure pertained to a spot on the ocean surface of the size of a few kilometres directly beneath the TOPEX/Poseidon satellite. By averaging the thousands of measurements collected by the satellite in the 10 days' time that it took to cover the global oceans, global mean sea level could be determined with a precision of the order of millimetres.

The TOPEX/Poseidon mission lasted about 14 years, until the January 2006. A comparison of surface wind speed and wave parameters derived from the TOPEX/Poseidon altimeter with ocean buoy data was made for the Indian Ocean by Kshatriya et al (2001).

4.4.2 Jason and OSTM

In December 2001, the Jason-1 satellite, another joint NASA-CNES venture, was launched as a replacement of the TOPEX/Poseidon mission and placed in the same orbit with similar instruments. Soon afterwards, in March 2002, ESA's ENVISAT satellite was launched with another radar altimeter in an orbit repeating at 35 days instead of 10 days for Jason and sampled the ocean on a smaller spatial scale, complementing the Jason data.

During the 5-year period when Jason-1 and TOPEX/Poseidon were both available, the two satellites were made to fly in tandem, providing twice the coverage of the sea surface. In the tandem mission, Jason-1 was positioned to move along the same ground track as that of TOPEX/Poseidon which was commanded to move onto a parallel ground track so that it was mid-way between two adjacent Jason ground tracks. The almost simultaneous measurements from the same altitude of the same sea surface locations by Jason and TOPEX/Poseidon altimeters enabled a rigorous comparison and correlation of their altimetry results.

Jason-1 functioned until 2013, much longer than expected. Its successor satellite, Jason-2, or the joint U. S. - Europe Ocean Surface Topography Mission (OSTM) was launched in June 2008. This marked a transition of satellite altimetry from research to an operational mode. The successor Jason-3 was launched in January 2016 and is operational. The primary instrument on Jason-3 is a radar altimeter that has an accuracy better than 3.3 cm.

4.5 References

Adler R. F. et al, 2003, "Status of TRMM monthly estimates of tropical precipitation", *Meteor. Monogr., Amer. Meteor. Soc.*, 29, 223 pp.

Asnani G. C. et al, 2000, "Patchy layered structure of tropical troposphere", *Indian J. Radio Space Physics*, 29, 182-191.

Asnani G. C., 2005, "Special features of tropics, monsoons", *Tropical Meteorology*, Vol. 1, Chapter 1, 16-37.

Bellouin N., Boucher O., Tanre D. and Dubovik O., 2003, "Aerosol absorption over the clear-sky oceans deduced from POLDER-1 and AERONET observations", *Geophys. Res. Lett.*, 30, 1748-1751.

Buriez J.-C. et al, 1997, "Cloud detection and derivation of cloud properties from POLDER", *Int. J. Remote Sens.*, 18, 2785–2813.

Cecil D. J., Zipser E. J. and Nesbitt S. W., 2002, "Reflectivity, ice scattering, and lightning characteristics of hurricane eyewalls and rainbands. Part I: quantitative description", *Mon. Wea. Rev.*, 130, 769-784.

Das P., George J. P., and Kumar S., 2016, "Verification of dust forecast over the Indian region with satellite and ground based observations", *International J. Remote Sensing,* 37, 5388-5411.

Deuze J. L. et al, 2000, "Estimate of the aerosol properties over the ocean with POLDER", *J. Geophys. Res.*, 105, 15329-15346.

Figa-Saldana J., et al, 2002, "The advanced scatterometer (ASCAT) on the meteorological operational (MetOp) platform: A follow on for European wind scatterometers", *Canadian J. Remote Sensing*, 28, 404-412.

Gelsthorpe R.V., Schied E. and Wilson J.J.W., 2000, "ASCAT-Metop's advanced scatterometer", *ESA Bulletin*, 102, 19-27.

Glecker, P. J. et al, 1995, "Cloud-radiative effects on implied oceanic energy transports as simulated by atmospheric general circulation models", *Geophys. Res. Lett.*, 22, 791–794.

Hazra A., Chaudhari H. S. and Dhakate A., 2015, "Evaluation of cloud properties in the NCEP CFSv2 model and its linkage with Indian summer monsoon", *Theoretical Applied Climatol.*, 69, 1-11.

Hirose M. and Nakamura K., 2005, "Spatial and diurnal variation of precipitation systems over Asia observed by the TRMM precipitation radar", *J. Geophys. Res.,* 110, D05106.

Kummerow C. et al, 1998, "The Tropical Rainfall Measuring Mission (TRMM) Sensor Package", *J. Atm. Ocean. Technol.*, 15, 809–817.

Kummerow C. et al, 2000, "The status of the Tropical Rainfall Measuring Mission (TRMM) after two years in orbit", *J. Appl. Meteor.*, 39, 1965–1982.

Kshatriya J., Sarkar A. and Raj Kumar, 2001, "Comparison of TOPEX/Poseidon altimeter derived wind speed and wave parameters with ocean buoy data in the north Indian Ocean", *Marine Geodesy*, 24, 131-138.

Liu W. T., 2002, "Progress in scatterometer application", *J. Oceanogr.,* 58, 121-136.

Murakami M., 1983, "Analysis of the deep convective activity over the western Pacific and southeast Asia, Part I: diurnal variation", *J. Meteor. Soc. Japan*, 61, 60-75.

Nesbitt S. W. and Zipser E. J., 2003, "The diurnal cycle of rainfall and convective intensity according to three years of TRMM measurements", *J. Climate*, 16, 1456-1475.

Raghavan S., 2003, *Radar Meteorology*, Kluwer Academic Publishers, 549 pp.

Rajeevan, M et al, 2013, "A study of vertical cloud structure of the Indian summer monsoon using CloudSat data", *Climate Dynamics,* 40, 637-650.

Rao P. B. et al, 1995, "Indian MST radar I –system description and sample vector wind measurements in ST mode", *Radio Sci.*, 30, 1125-1138.

Sarkar A., 2003, "Space based techniques for remote sensing of oceanic winds – a review", *Mausam*, 54, 111-120.

Srivastava R., Ramachandran S. and Rajesh T.A., 2016, "Aerosol mixing over an urban region: radiative effects", *Quarterly J. Royal Meteor. Soc.*, 142, 1732-1744.

Stephens G. L. et al, 2002, "The CloudSat mission and the A-Train: a new dimension of space-based observations of clouds and precipitation", *Bull. Amer. Meteor. Soc.*, 83, 1771–1790.

Stocker T. F. et al, 2001, "Physical climate processes and feedbacks", *Climate Change 2001 The Scientific Basis, IPCC WG I Third Assessment Report*, Chapter 7, 417-470.

Tao W.-K. et al, 2001, "Retrieved vertical profiles of latent heat release using TRMM rainfall products for February 1998", *J. Appl. Meteor.,* 40, 957-982.

Varma A. K., Liu G. and Noh Y. J., 2004, "Sub-pixel scale variability of rainfall and its application to mitigate the beam filling problem", *J. Geophys. Res.*, 109, D18210.

Wielicki B. A. Cess R. D., King M. D., Randall D. A. and Harrison E. F., 1995, "Mission to planet earth: role of clouds and radiation in climate", *Bull. Amer. Meteor. Soc.*, 76, 2125–2153.

Winker D. M., Pelon J. and McCormick M. P., 2003, "The CALIPSO mission: spaceborne lidar for observation of aerosols and clouds," *Proc. SPIE*, 4893, 1-11.

Winker D. M., Hunt W. and Hostetler C., 2004 "Status and performance of the CALIOP lidar", *Proc. SPIE*, 5575, 8-15.

Winker D. M. et al, 2009, "Overview of the CALIPSO mission and CALIOP data processing algorithms", *J. Atm. Oceanic Tech.*, 26, 2310-2323.

Yao M. S. and Del Genio A. D., 1999, "Effects of cloud parameterization on the simulation of climate changes in the GISS GCM", *J. Climate*, 12, 761-779.

Chapter 5

Missions Accomplished

Satellite missions involve years of planning, meticulous designing, difficult compromises, and elements of risk. Some satellites expected to have a short life actually live long. Some have a successful launch but come to a premature end. There are only a few that give a sense of pride and achievement. They are described in the subsequent sections of this chapter.

5.1 Megha-Tropiques

As discussed earlier, geostationary and polar orbiting satellites have their own relative merits and demerits in various respects. Geostationary satellites provide images every half-hour or even faster for small sectors. However, because of their great height, they are not suited for microwave remote sensing as the radiance reaching them is very weak. Orbiting satellites either in polar orbits or tropical orbits like TRMM, can provide a higher spatial resolution but they have a repeat cycle of one or two days and hence many atmospheric developments are lost between scans. The Megha-Tropiques satellite struck a trade-off among these conflicting requirements by placing passive microwave sensors in a low altitude orbit also having a low inclination with respect to the equatorial plane (Figure 5.1.1).

Megha-Tropiques was a joint India-France (ISRO-CNES) mission that was mooted in the late 1990s (Aguttes et al 2000). Both countries were to share the responsibility for development of payloads and ISRO was to use a PSLV launcher to place the satellite in a 867 km altitude and 20° inclination orbit. The expected mission life was five years. The project, however, could not make much headway for a long time because of a financial crunch. It was only in 2004 that the project was revived after India increased its contribution and overall costs were lowered. The satellite was launched as late as in October 2011 from ISRO's Satish Dhawan Space Centre in Sriharikota with a PSLV-C18 rocket. The mission ended in December 2021 having outlived itself by five more years.

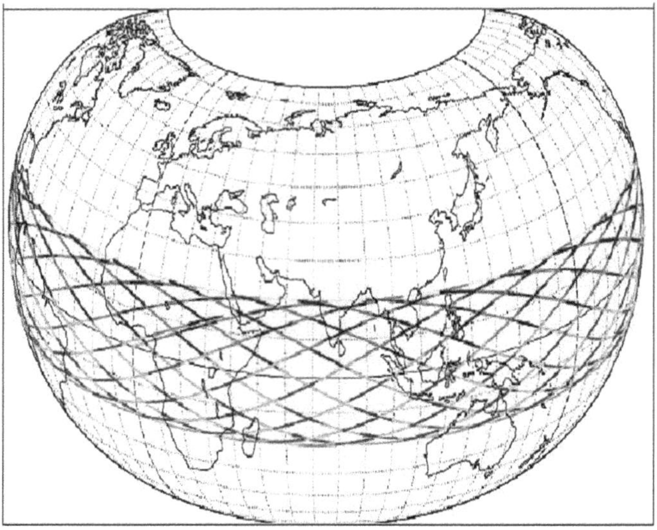

Figure 5.1.1 Megha-Tropiques orbit providing frequent imaging of the tropical belt.
(Source:https://www.eoportal.org/satellite-missions/megha-tropiques)

The main scientific objectives of the Megha-Tropiques mission were:

(a) To collect a long-term set of measurements with a good sampling and coverage over tropical latitudes to understand better the processes related to the water cycle and to evaluate their impact on the energy budget.

(b) To study the life cycle of tropical convective systems over ocean and continents, the environmental conditions for their formation and evolution, and the associated transport of water vapour.

(c) To study weather and climate events affecting tropical countries such as cyclones and systems producing heavy rainfall, and processes governing monsoon variability and drought.

Megha-Tropiques carried a rare combination of three state-of-art payloads, MADRAS, SAPHIR and ScaRaB designed for measurements of radiative fluxes, precipitation, humidity profiles and cloud properties, which are described below.

MADRAS (Microwave Analysis and Detection of Rain and Atmospheric Structures), was a passive imaging radiometer operating at five frequencies of 18.7, 23.8, 36.5, 89 and 157 GHz in both H and V polarizations except the 23.8 GHz which had only V polarization. Data from the first three channels had applications in the retrieval of rain over oceanic regions, liquid water content in clouds and vertical integrated water vapour. Their spatial resolution was 40 km. The 89 GHz channel was useful in retrieving convective rainfall over both land and ocean at a still better resolution of 10 km. The 157 GHz channel was to measure the concentration of ice particles in clouds at a resolution as high as 6 km. Unfortunately the microwave imager, MADRAS, stopped acquisition of scientific data in January 2013 due to a mechanical failure. During its 16 months of operation, it also experienced electrical issues making its usage difficult and delayed its validation.

SAPHIR (Sounder for Atmospheric Profiling of Humidity in the Inter-tropics by Radiometry) was a sounding instrument with 6 channels near the absorption band of water vapor at 183 GHz. These channels provided relatively narrow weighting functions from the surface to about 10 km, allowing retrieval of water vapour profiles in the cloud-free troposphere. The scanning was cross-track, up to an incidence angle of 50°. The ground resolution at nadir was 10 km.

ScaRaB (Scanner for Radiation Budget measurement) was the third Megha-Tropiques payload. It was an optical scanning radiometer devoted to the measurements of radiative fluxes at the top-of-atmosphere in the shortwave and longwave domain. ScaRaB had four channels: Sc1 - Visible 0.55 to 0.65 μ, Sc2 - Solar 0.2 to 4.0 μ, Sc3 - Total 0.2 to 100 μ, and Sc4 - IR window 10.5 to 12.5 μ. Sc2 and Sc3 were the main channels of the ScaRaB instrument. Longwave irradiance could be calculated from the difference between Sc3 and Sc2 measurements. The ground resolution varied from 40 km at nadir to 200 km on the edges. Images from Sc1 and Sc4 channels were to be used for scene identification and to provide the necessary compatibility with operational satellites like INSAT which had radiometers with similar spectral channels.

Although MADRAS had a short life, SAPHIR and SCARAB kept functioning nominally. It was therefore possible to retrieve a suite of geophysical products from these instruments, ranging from top of the atmosphere radiative fluxes to water vapour profiles and instantaneous rain rates. Some of these geophysical products have been merged with geostationary data to provide, for instance, daily accumulation of rainfall all over the intertropical region (Roca et al 2020). Megha-Tropiques also served as a part of the GPM constellation and contributed data to the overall global rainfall estimation process.

5.2 Tropical Rainfall Measuring Mission (TRMM)

The Tropical Rainfall Measuring Mission (TRMM) was a joint NASA/JAXA mission envisaged to fulfil the following science objectives (Kummerow et al 2000):

(a) to map the 4-dimensional structure of latent heat,

(b) understand the role of latent heat in the weather and climate of the tropics,
(c) obtain monthly average rainfall over the tropics and
(d) examine the diurnal cycle of tropical rainfall and the timing of heaviest rainfall.

In line with these objectives, the TRMM satellite was launched in November 1997 in a low earth orbit that viewed the tropical region many times a day. While TRMM was expected to have a life of three years, it went on working much longer. However, the satellite had to consume a lot of fuel to maintain its orbit because of its low altitude. So in August 2001, the satellite altitude was raised from 350 to 400 km in order to reduce the effects of atmospheric drag and further extend its useful life. Later on, other measures were adopted to optimize the battery use and carry on with the mission but by 2014 the fuel had run out. In April 2015, the TRMM mission officially came to an end, 17 years after it had been launched.

The TRMM data set built over this long period has been put to extensive use to quantify the relative contributions of convective and stratiform rain, develop a better understanding of the hydrological cycle, study the latent heat inputs to global energy and water cycles, study how tropical rainfall influences global circulation, and to improve modelling and prediction.

The TRMM mission, achieved remarkable scientific and operational milestones that reshaped our understanding of tropical precipitation and its role in global climate systems. TRMM was the first satellite to carry a Precipitation Radar (PR) which provided valuable insights into storm dynamics, vertical profiles of precipitation, and convective systems. It also carried the TRMM Microwave Imager (TMI) which helped estimate water vapour, cloud water, and rainfall intensity, and latent heat release in the tropics.

5.3 Global Precipitation Mission (GPM)

Much before the end of the TRMM mission, it was realized that it was necessary to work out its future continuity and also to widen the scope of radar-based measurement of precipitation from the tropics to the extra-tropics. A new international follow-on programme called the Global Precipitation Measurement Mission (GPM) was therefore planned (Flaming 2005) and it became operational in 2014.

The GPM mission had four broad and basic elements:

(a) A core spacecraft carrying a Dual-frequency Precipitation Radar (DPR) and a GPM Microwave Imager (GMI) to make the primary radar-based rainfall measurements and also serve as a calibration reference for other instruments operating under GPM,

(b) A large constellation of satellites to be provided by various space agencies, equipped with passive microwave radiometers,

(c) A ground validation programme for verification with reference to ground truth over different weather regimes, and

(d) A data processing system to collect, process and disseminate the vast amount of data gathered by the core satellite and other members of the GPM constellation.

The GPM core spacecraft was launched in February 2014 with a Japanese H-IIA rocket into a 65° inclination orbit at a 400 km altitude. This orbit excludes the latitudes poleward of 65° N and S, but covers all the regions where precipitation occurs in the form of rain. As was the case with TRMM, JAXA developed and provided the DPR and the launch vehicle while NASA provided the passive microwave radiometer, the satellite bus, and the ground control segment. While TRMM measured precipitation in the tropics, the GPM Core Observatory expands the coverage area from the Arctic Circle to the Antarctic Circle.

GPM is a fine example of close and fruitful international collaboration among space agencies. Besides the core satellite, the GPM constellation is made up of several satellites of partner agencies. Megha-Tropiques satellite was a part of the GPM constellation.

The GPM mission was originally designed with a nominal life of 3 years and a goal of 5 years. However, it has far exceeded expectations and in spite of problems related to the reaction wheels, the core satellite remains operational in an extended phase. The end-of-life is presently projected as September 2026.

Cyclone Tauktae developed over the Arabian Sea on 13 May 2021, intensified into an Extremely Severe Cyclonic Storm, or a Category 4 equivalent, on 17 May and made landfall on the Gujarat coast the same day. It was one of the strongest tropical cyclones to ever affect the west coast of India. An active phase of the Madden-Julian Oscillation (MJO) was prevalent at that time over the Indian Ocean and was accompanied by a westerly wind burst just north of the equator.

Figure 5.3.1 shows the GPM measured precipitation of the order of 250 mm over the Arabian Sea when Tauktae was close to Mumbai. Figure 5.3.2 shows the GPM week-long rain accumulation off the entire western coastline of India.

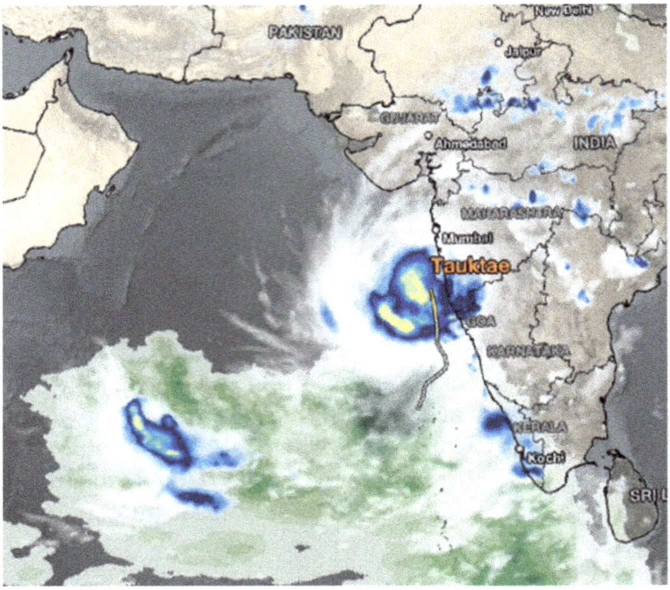

Figure 5.3.1 GPM rainfall around cyclone Taukte over the Arabian Sea on 17 May 2021. (Source: https://gpm.nasa.gov/applications/weather/imerg)

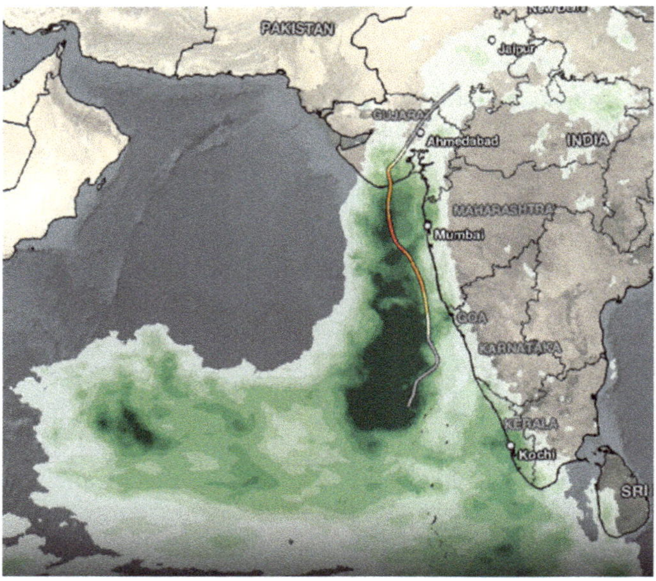

Figure 5.3.2 GPM rainfall accumulation around cyclone Taukte's track. (Source: https://gpm.nasa.gov/applications/weather/imerg)

5.4 A-Train

The term "formation flying" is commonly used to describe the disciplined and orderly flying of birds as well as aircrafts that results in the display of a geometrical pattern in the sky. In the early years of this century, this concept was extended to satellites. Satellites in a formation flying mode, were made to fly in matched orbits, keeping a safe distance from each other and maintaining a predetermined relative geometry.

The A-Train concept (Anderson et al 2005) provided for the first time, a synergy among different sensors, overlapping footprints and complementary capabilities in terms of retrieved quantities, sensitivity, resolution, coverage, and joint product retrievals. It made it possible to obtain an ensemble of near-simultaneous measurements of aerosols, clouds, temperature, relative humidity and radiative fluxes over the globe during all seasons.

The CALIPSO and CloudSat satellites had highly complementary functions and they were the first to fly together to provide totally new 3-dimensional perspectives of clouds and aerosols. Their lidar and radar footprints had a direct overlap more than 70 per cent of the time and within 1.5 km of each other at all times (Stephens et al 2002). Later on, more satellites were added to the formation.

In the original A-Train configuration design, all the satellites had a nominal orbital altitude of 705 km and inclination of 98° with a repeat cycle of 16 days. The Aqua satellite led the train with an equatorial crossing time of about 1:30 pm. CloudSat and CALIPSO trailed behind it by just a couple of minutes and the two satellites themselves were separated from each other by just 10-15 sec (Vincent et al 2004). PARASOL lagged behind CALIPSO by 1 min. Aura trailed Aqua by about 15 min but it crossed the equator 8 min behind Aqua due to its different orbital track which allowed for synergy with Aqua. The Orbiting Carbon Observatory (OCO) preceded Aqua by 15 min.

Actually, however, the 2006 configuration has had to be modified from time to time. OCO failed in 2009 and was replaced by OCO-2 in 2014. In 2012, the Japanese satellite, Global Change Observation Mission - Water (GCOM-W1) was added to A-Train, but in 2013, PARASOL had to be dropped.

As of 2014, the A-Train had six active satellites. OCO-2 was the lead spacecraft in formation, GCOM-W1 followed it by 11 min, and Aqua ran 4 minutes behind GCOM-W1. CALIPSO ran 2.5 min behind Aqua and CloudSat followed CALIPSO by 15 sec. Aura, lagged behind Aqua by 15 min (Figure 5.4.1)

CloudSat and CALIPSO ended their missions in 2023. As of September 2025, both Aqua and Aura continue to collect science data; however, they have slowly drifted out of the constellation.

Figure 5.4.1 The A-Train concept.
(Source: NASA web site atrain.nasa.gov/images)

5.5 WindSat

The principle of polarimetric radiometry is that while the visible solar radiation coming to the earth from the sun is not polarized, certain physical processes on earth result in generating polarized light. If satellites are able to measure this polarization, then properties of clouds and aerosols and other atmospheric parameters which caused the polarization can be estimated. The principle of polarimetric radiometry was successfully used in the POLDER instrument which flew on three different satellites, ADEOS-I, ADEOS-II and PARASOL (Section 4.2.1).

WindSat/Coriolis was a sun-synchronous satellite in low earth polar orbit that was launched by the U.S. Department of Defense in 2003. Its primary aim was to demonstrate the capability of polarimetric microwave radiometry to measure the ocean surface wind vector from space (Gaiser et al 2004). While WindSat had a design life of only three years, data continued to be received until 2020.

The WindSat polarimetric radiometer operated in discrete bands at 6.8, 10.7, 18.7, 23.8, and 37 GHz. The 10.7, 18.7 and 37 GHz channels were fully polarimetric. The 6.8 GHz channel had dual polarization and was used to remove environmental noise due to variations in SST. The 23.8 GHz channel also had dual polarization and was used to correct for the effects of atmospheric attenuation on signals from the ocean surface.

This measured both vertical and horizontal polarizations of the microwave radiation reaching the satellite, and in addition measured the cross-correlation between the two. The cross-correlation terms were derived from the microwave radiances a ertical, horizontal, and 45° left hand circular and right hand circular polariz ns. The Stokes vector provided a full characterization of the spectral sign e of the ocean surface in the microwave region and yielded all the inde dent information needed to uniquely determine the wind direction over th an (Lahtinen et al 2003).

A compar f WindSat sea surface winds with those derived from active and pa icrowave instruments carried by QuikSCAT and ASCAT satellites fo rs 2003 to 2010 was made over the Mediterranean Sea by De Biasio 013). The main results of the comparison showed up the unsuitabi e WindSat data for coastal applications, and a general degrada ata quality with time and some systematic errors. Wind retriev olarimetric radiometers are best made when there is no rain, but t also be used to obtain additional information on parameters suc face temperature, rain rate, ice and snow characteristics, water v l moisture (Li et al 2010, Parinussa et al 2012).

ernational Satellite Cloud Climatology Project CCP)

nd of the 1970s, an operational system of polar orbiting and nary satellites had been put into place, and a massive amount of data g collected by the satellite agencies. Until this time, the main objective atellite operators had in view was the use of imagery for short range ather forecasting and the retrieval of parameters in a form that weather forecasters were accustomed to see plotted on their synoptic charts. By the beginning of the 1980s, however, there was a paradigm change in this thinking. It was realized that carefully analyzed and processed satellite data and products had great value from the point of view of long-term climate monitoring, numerical modelling and climate change studies. At that point of time, only passive sensors were in vogue, and it was felt that whatever information that could be gathered from passive sensors about the nature and distribution of clouds and precipitation should be systematically used and processed. Accordingly, two international programmes were conceived.

Compared to the methods adopted in conventional climatology, the generation of long-term satellite data sets for climate purposes is a very demanding exercise. First of all, there is a need for a very strict quality control of the satellite data and the derived products. Sensors on board satellites show a degradation with time, which has to be accounted for and the data corrected accordingly. Unlike meteorological instruments on the ground, satellite sensors are constantly improving and changing. Different types of sensors are used on different satellites. Hence a lot of effort has to be put in to make the data sets compatible across different sensors and comparable throughout the time series. Merger of satellite derived parameters an[d] their observations on the ground is another difficult problem as the two ma[y n]ot necessarily have the same physical meaning. At the same time, there is alternative to the necessity of evaluating and validating the satellite proc[ducts] against what is available as ground truth.

The International Satellite Cloud Climatology Proj[ect] (ISCCP) was established as part of the World Climate Research Program (WCRP) in 1982 (Schiffer et al 1983) with several specific objectives, the prim[ary] being the production of a global, calibrated and normalized data set ... infrared radiances along with basic information on the radiative ... e and the atmosphere, from which cloud parameters could be derived. A ... s of was put in towards developing techniques and algorithms. Oper... ort collection for the ISCCP began in July 1983 and it was a complex ...

The primary product of ISCCP was the B3 data set beginning i[n] 1983 which was a collection of VIS and IR radiance images from a[ll] operational weather satellites in a common format and normalize[d to a] standard reference calibration. There were several other ISCCP data set[s at] a variety of space and time resolutions and processed differently to s[erve] different application areas (Rossow et al 1991).

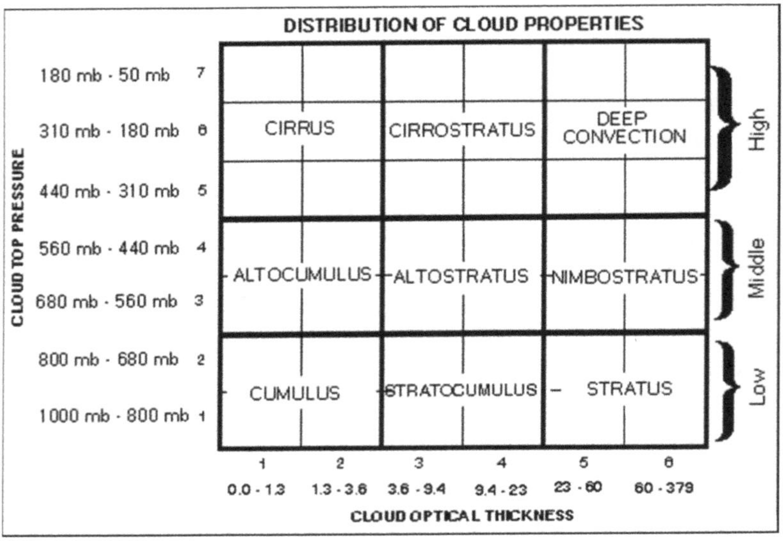

Figure 5.6.1 ISCCP cloud classification.
(Source: https://isccp.giss.nasa.gov/)

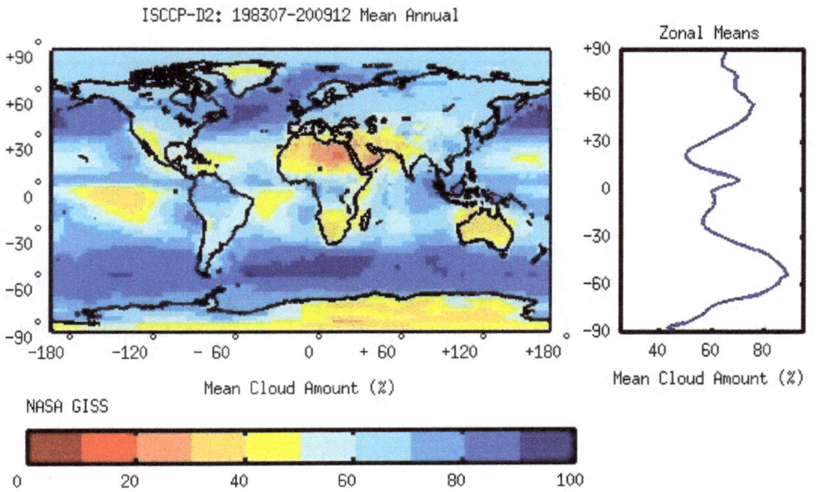

Figure 5.6.2 ISCCP global mean cloud amount and zonal means. (Source: https://isccp.giss.nasa.gov/)

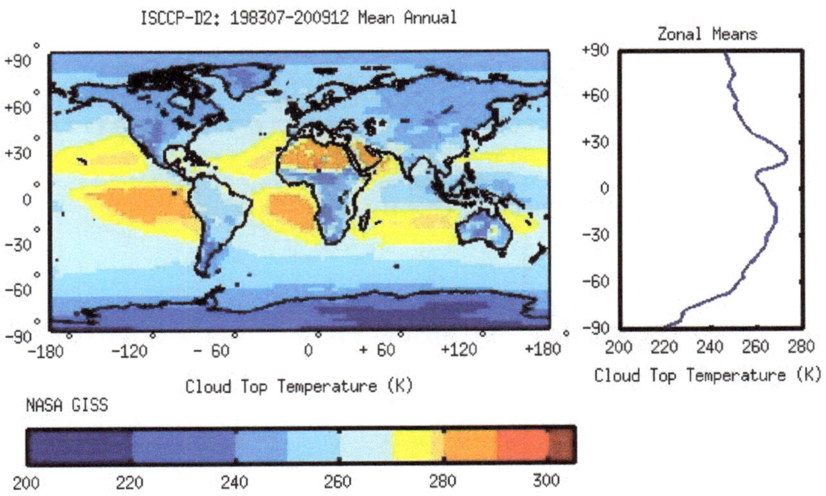

Figure 5.6.3 ISCCP global mean cloud top temperature and zonal means. (Source: https://isccp.giss.nasa.gov/)

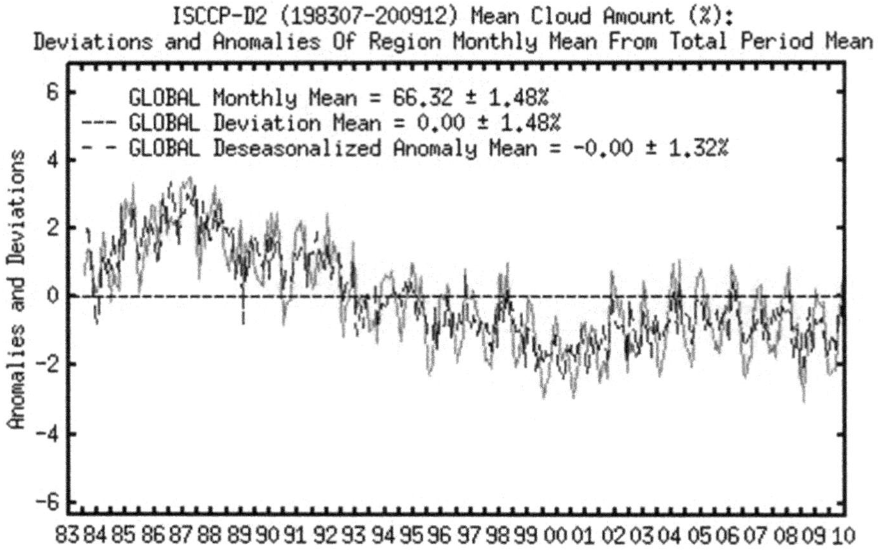

Figure 5.6.4 ISCCP time series of global cloud amount monthly anomalies. (Source: NASA web site isccp.giss.nasa.gov/zD2BASICS)

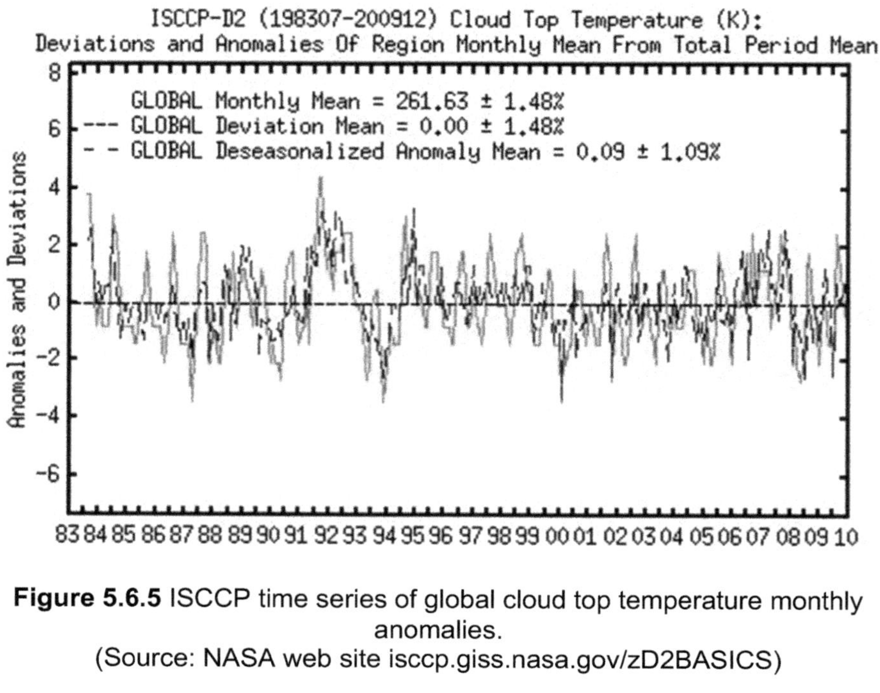

Figure 5.6.5 ISCCP time series of global cloud top temperature monthly anomalies.
(Source: NASA web site isccp.giss.nasa.gov/zD2BASICS)

The cloud classification scheme adopted by ISCCP was that of two-dimensional frequency distributions or histograms of the cloud top pressure and visible optical thickness of which there were nine different combinations (Figure 5.6.1). The middle and low level cloud types could also be separated into liquid and ice clouds. Each category was given a name that conforms to conventional cloud types (Hahn et al 2001).

A significant result of the 26-year ISCCP analysis from 1983 to 2009 was that the global annual mean cloud amount is about 67%. The mean annual Southern Hemisphere cloud amount is 6% greater than the mean Northern Hemisphere cloud amount because of differences in hemispheric land fractions. The mean annual ocean cloud amount is about 23% larger than the land cloud amount. The mid-latitude zones 30°-60° have the largest mean cloud amount, about 10% larger than the global mean value. Figures 5.6.2 and 5.6.3 show the 1983-2009 averages of cloud amount and cloud top temperature respectively across the globe and also their zonal means. The global mean cloud amount has shown an increase in the initial period of the ISCCP data series and then a decrease but the cloud top temperature does not show such a change (Figures 5.6.4 and 5.6.5).

Although ISCCP has met its original mission objectives in 2009, it is still engaged in a limited way with the compilation of what is termed as the Cloud Properties-ISCCP H-Series Climate Data Record (CDR). The CDR includes all upwelling and downwelling, shortwave and longwave radiative fluxes at the top of the atmosphere, at three levels in the atmosphere, and at the surface, using a complete radiative transfer model (Zhang and Rossow 2023). The CDR currently extends from 1983 to 2018. ISCCP now focuses on the distribution and variation of cloud radiative properties to improve the understanding and modelling of the way clouds affect climate. ISCCP thus continues to help in cloud parameterization in climate models, global energy budget and hydrological cycle studies and long term climate trend analysis through consistent satellite derived cloud data sets.

5.7 Global Precipitation Climatology Project (GPCP)

The ISCCP initiative of 1982 was closely followed by the launch of another major international undertaking that would complement the ISCCP. The Global Precipitation Climatology Project (GPCP) was established by the WCRP in 1986 to address the problem of quantifying the long term distribution of global precipitation. The initial goal was that of deriving monthly mean precipitation data on a 2.5° × 2.5° latitude-longitude grid. Later, the data sets were expanded to include a 2.5° × 2.5° pentad data set (Xie et al 2003) and a 1° × 1° daily data set (Huffman et al 2001). The GPCP objective was to make these data sets available to validate general circulation and climate models, study the global hydrological cycle and analyze the variability of the global climate system.

GPCP began producing monthly mean precipitation estimates from 1979 onwards. The general approach was to combine the precipitation information available from several sources: microwave estimates from DMSP SSM/I, IR estimates from geostationary satellites, and IR estimates from NOAA AVHRR and TOVS, besides data from more than 6000 rain gauges. GPCP has promoted the development of an analysis procedure for blending the various estimates together to produce the necessary global gridded precipitation fields (Adler et al 2003, Huffman et al 2009). The GPCP Version 2.2 Combined Precipitation Data Set has a total of 27 fields.

GPCP culminated in the compilation of the global and seasonal distributions of precipitation which have a sound foundation and authenticity. Figures 5.7.1 and 5.7.2 show the GPCP mean monthly global rainfall distributions for January and July respectively. In the July climatology, the Bay of Bengal, northeast India and the west coast of India stand out as the zones of highest rainfall in the world (more than 20 mm/day), matched only by the eastern Pacific Ocean. At the same time, the southwestern Arabian Sea is clearly seen as an almost rainless region. In the January climatology, rainfall does not generally exceed 10 mm/day, and the heavy rainfall belt is seen to be aligned with the ITCZ over the southern hemisphere.

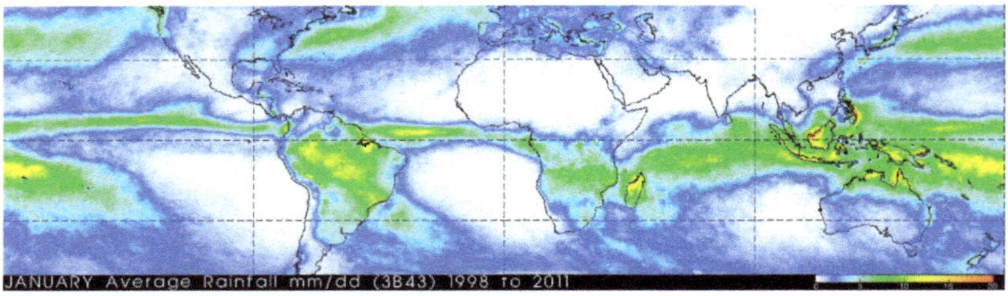

Figure 5.7.1 GPCP mean monthly global rainfall distribution for January. (Source: https://www.psl.noaa.gov/data/gridded/data.gpcp.html)

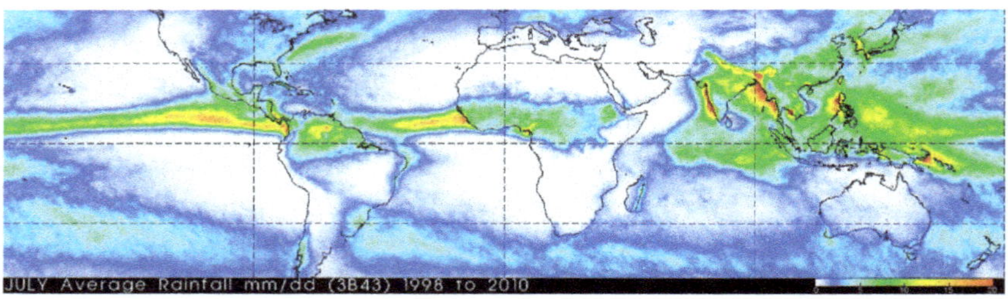

Figure 5.7.2 GPCP mean monthly global rainfall distribution for July. (Source: https://www.psl.noaa.gov/data/gridded/data.gpcp.html)

Although GPCP started processing precipitation data in 1979 and has produced voluminous data sets, it continues to compile and update them even now. Recent studies using GPCP data have revealed significant changes in global precipitation patterns over time (Huffman et al 2023).

5.8 References

Adler R.F. et al, 2003, "Version 2 Global Precipitation Climatology Project (GPCP) monthly precipitation analysis (1979-present)", *J. Hydrometeor.*, 4, 1147-1167.

Aguttes J. P. et al, 2000, "Megha-Tropiques, a satellite for studying the water cycle and energy exchanges in the tropics", *Geoscience and Remote Sensing Symposium, 2000. Proceedings. IGARSS 2000. IEEE 2000 International*, 7, 3042-3044.

Anderson T. L. et al, 2005, "An A-Train strategy for quantifying direct climate forcing by anthropogenic aerosols", *Bulletin American Meteor. Soc.*, 86. 1795-1810.

De Biasio, F. and Zecchetto S., 2013, "Comparison of sea surface winds derived from active and passive microwaves instruments on the Mediterranean Sea". *EGU General Assembly Conference Abstracts*, 10157.

Flaming G. M., 2005, "Global precipitation measurement update. *Proc.. 2005 IEEE International Geoscience and Remote Sensing Symposium,* IGARSS'05. 1, 4-5.

Gaiser P. W. et al, 2004, "The WindSat spaceborne polarimetric microwave radiometer: sensor description and early orbit performance", *IEEE Transactions on Geoscience and Remote Sensing,* 42, 2347-2361.

Hahn C. J., Rossow W. B. and Warren S. G., 2001, "ISCCP cloud properties associated with standard cloud types identified in individual surface observations", *J. Climate*, 14, 11-28.

Huffman G. J. et al, 2001, "Global precipitation at one-degree daily resolution from multi-satellite observations", *J. Hydrometeor.*, 2, 36-50.

Huffman G. J., Adler R. F., Bolvin D. T. and Gu G., 2009, "Improving the Global Precipitation Record: GPCP Version 2.1", *Geophys. Res. Lett.*, 36, L17808, doi:10.1029/2009GL040000

Huffman, G, J. et al, 2023, "The new version 3.2 Global Precipitation Climatology Project (GPCP) monthly and daily precipitation products", *J. Climate,* 36. 7635-7655.

Lahtinen, J., 2003, "Fully polarimetric radiometer system for airborne remote sensing", *Helsinki University of Technology.*

Li L. et al, 2010, "WindSat global soil moisture retrieval and validation", *IEEE Trans. Geoscience Remote Sensing,* 48: 2224-2241.

Parinussa R. M., Holmes T. R. H. and de Jeu R. A. M., 2012, "Soil Moisture retrievals from the WindSat spaceborne polarimetric microwave radiometer", *IEEE Trans. Geosci. Remote Sensing*, 50, 2683–2694.

Roca R. et al, 2020, "The Megha-Tropiques mission after seven years in space", *Satellite Precipitation Measurement: Volume 1,* Springer International Publishing, 2020. 45-62.

Rossow, W.B. and Schiffer R. A., 1991, “ISCCP cloud data products”, *Bull. Amer. Meteor. Soc.*, 72, 2-20.

Schiffer R. A. and Rossow W. B., 1983, “The International Satellite Cloud Climatology Project (ISCCP): the first project of the World Climate Research Programme”, *Bull. Amer. Meteor. Soc.*, 64.

Stephens G. L. et al. 2002, "The CloudSat mission and the A-Train: a new dimension of space-based observations of clouds and precipitation." *Bulletin American Meteor. Soc.*, 83, 1771-1790.

Vincent M. A. and Salcedo C., 2004, "The insertion of CloudSat and CALIPSO into the A-Train constellation", *Adv. Astronautical Sci.,* 116, 1401-1418.

Xie P. et al, 2003, “GPCP pentad precipitation analyses: an experimental dataset based on gauge observations and satellite estimates”. *J. Climate*, 16, 2197-2214.

Zhang Y. and Rossow W. B., 2023, “Global radiative flux profile dataset: revised and extended”, *J. Geophys. Res. Atmos.*, 128, 2022JD037340,

Chapter 6

Current Global Configuration of Operational Satellites

Although several options are now available for satellite orbits, the operational meteorological satellites of all countries and agencies are still either geostationary or polar orbiting. Considering that a geostationary satellite located above the equator at a height of 36,000 km can view about 65° x 65° latitude/longitude area of the earth's disc, a system of five or six such satellites separated equidistantly would suffice to cover the earth. Likewise, two polar orbiting satellites, one crossing the equator in the morning and another in the afternoon would give adequate global coverage.

6.1 Global Configuration of Meteorological Satellites

Presently, the situation is far more comfortable compared to this nominal scenario, with as many as 18 geostationary satellites of 7 countries or agencies being fully operational or in backup readiness as of 1 September 2025 (Table 6.1.1). Eumetsat has five satellites, China has four, U.S. three, Russia and India two each, and Japan and South Korea one each. The global meteorological satellite configuration does keep changing with time, however, as old satellites get phased out and new ones take their place. Even the present number of polar orbiting meteorological satellites far exceeds the minimal requirement.

As of 1 September 2025, there were 9 satellites of this category in operation (Table 6.1.2). U.S. and China have three each, Russia has two, and Eumetsat has one. India does not have a polar orbiting meteorological satellite yet. Here again, the global network is liable to change in future.

Table 6.1.1 Global Configuration of Geostationary Meteorological Satellites as of 1 September 2025.

Longitude	Name of Satellite	Year of Launch	Operating Country/ Agency
0 ºE	Meteosat-13	2025	Eumetsat
3.5 ºE	Meteisat-10	2012	Eumetsat
9.5 ºE	Meteosat-11	2015	Eumetsat
45.5 ºE	Meteosat-9	2005	Eumetsat

Table 6.1.1 *Contd....*

Longitude	Name of Satellite	Year of Launch	Operating Country/ Agency
74 °E	INSAT-3DR	2016	India
76 °E	Electro-L N3	2019	Russia
79 °E	FY-2H	2018	China
82 °E	INSAT-3DS	2024	India
99.5 °E	FY-2G	2014	China
105 °E	FY-4B	2021	China
123. °E	FY-4A	2016	China
128.2 °E	GEO-KOMPSAT-2B	2020	S. Korea
140.7 °E	Himawari-9	2016	Japan
165.8 °E	Electro-L N4	2023	Russia
137 °W	GOES-18 (West)	2022	U.S.A.
104.7 °W	GOES-17 (Backup)	2018	U.S.A.
75.2 °W	GOES-19 (East)	2024	U.S.A.
0.3 °W	Meteosat-12	2022	Eumetsat

(Source: https://space.oscar.wmo.int/satellitestatuses/status)

Table 6.1.2 Global Configuration of Polar Orbiting Meteorological Satellites as of 1 September 2025.

Equator Crossing Time	Name of Satellite	Year of Launch	Operating Country/ Agency
Early Morning	Suomi NPP	2011	U.S.A.
	FY-3E	2021	China
Morning	Meteor-M N2-3	2023	Russia
	MetOp-C	2018	Eumetsat
	FY-3F	2023	China
	NOAA-20	2017	U.S.A.
Afternoon	FY-3D	2017	China
	Meteor-M N2-4	2024	Russia
	NOAA-21	2022	U.S.A.

(Source: https://space.oscar.wmo.int/satellitestatuses/status)

6.2 U.S. – GOES, NOAA and Suomi NPP

In the past, geostationary satellites faced two severe limitations. Firstly, the radiometers were not sensitive enough to measure the radiances from the earth and atmosphere which weaken considerably by the time they reach the great height of the satellites. Secondly, the instruments had a comparatively low signal-to-noise ratio. With recent advances in sensor technology, however, these limitations have largely been overcome. The current generation of the U.S. Geostationary Environmental Operational Satellite (GOES), named as GOES-R, is very advanced compared to its predecessors (Schmit et al 2005, 2017).

GOES-R is a three-axes stabilized geostationary satellite. The Advanced Baseline Imager (ABI) on GOES-R is therefore a multi-spectral, scanning radiometer with alternate east-west and west-east scans. ABI operates in as many as 16 spectral bands including 2 VIS, 4 NIR and 10 IR channels. The ground resolution of ABI is 0.5 km for the 0.64 μ VIS band, 2 km for the IR bands and 1 km for the other VIS and NIR channels. GOES-R can operate in several scan modes, ranging from full disc in 15 min, contiguous U.S. every 5 min, to mesoscale mode over a 1000 x 1000 km box in 30 sec.

GOES-R also has extreme ultraviolet and x-ray irradiance sensors, and a solar ultraviolet imager. In addition to these payloads, it has a Geostationary Lightning Mapper (GLM). This is a single-channel, near-infrared optical transient CCD detector that can detect the momentary changes in an optical scene, indicating the presence of lightning. GLM measures total lightning activity continuously over the Americas and adjacent ocean regions with near uniform spatial resolution of approximately 10 km. On the negative side, previous GOES satellites had a sounding instrument named Visible Infrared Spin-Scan Radiometer Atmospheric Sounder (VAS) but it has been discontinued with the GOES-R series.

The Suomi NPP (Suomi National Polar-orbiting Partnership) satellite was launched in October 2011 and named after the late Verner E. Suomi who is regarded widely as the father of satellite meteorology. Suomi had done pioneering work in meteorological remote sensing beginning with orbiting satellites like Explorer-7 in 1959 and geostationary satellites such as ATS-1 in 1966. He invented the spin-scan camera that later became the basic onboard instrument of numerous meteorological satellites. Suomi also contributed significantly to the design of many U.S. space missions to other planets of the solar system. He devoted almost his entire career to the University of Wisconsin at Madison, where in 1965 he founded the Space Science and Engineering Centre. Suomi died in 1995 at the age of 79.

NOAA-20 was launched in November 2017, and NOAA-21 in December 2022. Both these satellites, as well as the SuomiNPP, have the following key instruments:

(a) Visible Infrared Imaging Radiometer Suite (VIIRS), a 22-band VIS and IR radiometer for weather monitoring

(b) Advanced Technology Microwave Sounder (ATMS), a microwave radiometer for global temperature and moisture profiling

(c) Cross-track Infrared Sounder (CrIS), a Michelson interferometer to monitor moisture and pressure

(d) Ozone Mapping and Profiler Suite (OMPS), a group of imaging spectrometers to measure ozone levels, especially near the poles

(e) Clouds and the Earth's Radiant Energy System (CERES), a radiometer to detect the solar radiation reflected by the earth and the thermal radiation emitted by it

Lee et al (2006, 2010) and Miller et al (2006) have given complete details of the VIIRS payload, including its characteristic features and the pixel dimensions in the along- and cross-track directions. Out of the total 22 VIIRS channels, 9 are in VIS and NIR (0.4-0.9 μ), 8 are in SWIR and MWIR (1-4 μ), 4 in the TIR (8-12 μ), and one in the low-light VIS. While VIIRS is by all means an ambitious instrument from an operational viewpoint, its primary shortcomings are the absence of the now familiar WV channel at 6.7 μ.

As far as the resolution is concerned, out of the 22 channels, 7 channels to be used for ocean colour and SST applications have a ground resolution of 0.742 km × 0.259 km, 9 channels to be used for deriving cloud properties and SST have a resolution of 0.742 × 0.776 km, and 5 imaging channels have a resolution of 0.371 × 0.387 km.

6.3 Europe – Meteosat and MetOp

Meteosat-9, -10 and -11 are the remainders of the Meteosat Second Generation (MSG) series and Meteosat-12 and -13 are the frontrunners of the Meteosat Third Generation (MTG) series. These five European satellites together cover the continents of Europe and Africa and parts of the Indian Ocean.

One MSG satellite provides full disc imagery of the European and African continents and parts of the Atlantic and Indian oceans every 15 minutes. Another MSG satellite delivers more frequent images every five minutes (rapid scanning service) over Europe only, to further support warnings of localized high-impact weather such as thunderstorms and flooding.

The third operational MSG satellite is situated over the Indian Ocean and contributes to the Indian Ocean Data Coverage (IODC) service, from where it provides data crucial for the safety of the people of island nations and eastern Africa. It is also important for improved weather forecasting in Central and Eastern Europe, the Middle East, and Central Asia. The two primary MSG instruments are SEVIRI (Spinning Enhanced Visible and Infrared Imager) and GERB (Geostationary Earth Radiation Budget).

SEVIRI is a multispectral radiometer operating in 12 spectral channels, 8 in TIR, 3 in VIS and NIR, and 1 high-resolution VIS. The imaging cycle repeats every 15 min. GERB is a broadband VIS-IR radiometer covering the solar spectrum from 0.32 to 4.0 μ and the terrestrial spectrum from 0.32 to 30 μ. The function is to measures the earth's radiation budget components like reflected solar radiation and emitted thermal radiation.

The overall Meteosat Third Generation (MTG) programme comprises six satellites, four imaging satellites (MTG-I) and two sounding satellites (MTG-S), all of them geostationary and three-axes stabilized. The first MTG-I satellite, Meteosat-12, was launched in December 2022 and is now operational in geostationary orbit covering Europe and Africa and surrounding oceans. The first MTG-S satellite, Meteosat-13, was launched on 1 July 2025.

The MTG-I satellites carry an imaging payload comprising the Flexible Combined Imager (FCI) and the Lightning Imager. The MTG-S satellites have the Infrared Sounder (IRS) and the Ultra-violet, Visible and Near-infrared Sounder (UVN). The sounder will be one of the key innovations in the Meteosat programme, which has so far had only imagers. Design features of the MTG instruments have been discussed by Schmetz et al (2012).

FCI provides continuity with the earlier SEVIRI instrument but it has 16 channels against SEVIRI's 12. The spatial resolution is 1–2 km in the standard mode and 0.5 km in the fast Imagery mode. The Infrared Sounder (IRS) is a new feature for European geostationary satellites. It is a hyperspectral sounder using 1,960 MWIR and LWIR spectral channels per dwell. The soundings are produced every 30 min.

MetOp-B was launched in September 2012 and MetOp-C in November 2018. These satellites are in sun-synchronous orbits at an altitude of 824 km with an orbital inclination of 98.7°. Their orbital period of these satellites is 101.4 minutes with equator crossing time of 9:30 am. Currently MetOp-C is in operation with MetOp-B as a backup. They carry the following instruments:

(a) Infrared Atmospheric Sounding Interferometer (IASI), which obtains humidity and atmospheric temperature profiles in the troposphere and lower stratosphere

(b) Microwave Humidity Sounder (MHS), which makes passive microwave measurements for deriving atmospheric humidity profiles

(c) Global Navigation Satellite System Receiver for Atmospheric Sounding (GRAS), in which a GPS receiver is used to derive atmospheric temperature and humidity profiles through the process of GPS radio occultation

(d) Advanced Scatterometer (ASCAT), an enhanced follow-on to the scatterometers flown on the ERS-1 and ERS-2 satellites

(e) Global Ozone Monitoring Experiment-2 (GOME-2), for measurements of ozone and other atmospheric constituents

IASI (Cohen et al 2006) is a Michelson interferometer working in 8461 spectral channels and measures temperature, water vapour and trace gases. The Integrated Imaging System (IIS) is a part of IASI and it consists of a broadband radiometer working at 10-12 μ at high spatial resolution. The field of view covers 64 x 64 pixels and provides information in the focal plane of the sounder, allowing precise co-registration of imager and sounder pixels. The IASI sounding is also synchronized with the microwave sounding. MHS is a new instrument that replicates the AMSU-B payload flying on NOAA satellites. MetOps also carry several other instruments that provide commonality with the NOAA orbiting satellites.

The MetOp Second Generation mission is an advancement over the current MetOp, and the first satellite MetOp-SG-A1 of the new series was launched 13 August 2025. It will be followed by its twin satellite, MetOp-SG-B1 in 2026. Three such pairs of satellites are planned for the future over the next 20 years. While some instruments on each MetOp-SG satellite are improved versions of the ones on the first MetOp series, others are entirely new.

The MetOp-SG-A1 satellite carries a next-generation infrared atmospheric sounder, a microwave sounder, a multispectral imaging radiometer, a novel multi-viewing, multi-channel, multi-polarization imager, a radio occultation sounder, and the European Commission's Copernicus Sentinel-5 spectrometer (Sentinel-5A), a total of six instruments.

The Sentinel-5A is one of the MetOp-SG-A1's key instruments. It is an advanced imaging spectrometer designed to provide daily global data on major air pollutants, key climate indicators, and the stratospheric ozone layer that protects the Earth from harmful ultraviolet radiation.

6.4 India – INSAT

INSAT-3DR located at 74°E and INSAT-3DS at 82°E are India's geostationary satellites that are currently operational. INSAT-3DS full disc images are available at 0000, 0030, 0100… UTC while INSAT-3DR full disc images are available at 0015, 0045, 0115… UTC. In effect, images of the Asian and Indian Ocean region get updated every 15 min. In severe weather situations, rapid scan mode is used to get images of a selected area every 5 minutes.

Both INSAT-3DR and INSAT-3DS have on board a 6-Channel Imager and a 19-Channel Sounder that enables the retrieval of vertical profiles of temperature and humidity at a 10 km ground resolution every three hours (Tables 6.4.1 and 6.4.2).

Table 6.4.1 INSAT-3D 6-Channel Imager.

Channel	Wavelength Range (μ)	Ground Resolution (km)
VIS	0.55-0.75	1
SWIR	1.55-1.70	1
MWIR	3.80-4.00	4
WV	6.50-7.10	8
TIR 1	10.3-11.3	4
TIR 2	11.5-12.5	4

Table 6.4.2 INSAT-3D 19-Channel Sounder.

Channel Number	Central Wavelength (μ)	Main Absorbing Gas
1	14.71	Carbon dioxide
2	14.37	
3	14.06	
4	13.96	
5	13.37	
6	12.66	Water vapour
7	12.02	
8	11.03	Window
9	9.71	Ozone
10	7.43	Water vapour
11	7.02	
12	6.51	
13	4.57	Nitrous oxide
14	4.52	
15	4.45	Carbon dioxide
16	4.13	
17	3.98	Window
18	3.74	
19	0.695	Visible

6.5 Japan – Himawari

All Japanese meteorological satellites are named as "Himawari" meaning "sunflower". The geostationary satellite Himawari-9 carries an instrument called the Advanced Himawari Imager (AHI). This is quite similar to the ABI on the GOES satellites. AHI has a total of 16 bands, 6 in the VIS and NIR, and

10 TIR bands. The spatial resolution 2 km, AHI scans the full disc every 10 min, and can smaller regions in 2.5 min.

6.6 China – FY

All Chinese meteorological satellites are named as "Fengyun" (FY) meaning "Wind and Cloud". Several satellites from the FY series are currently operational. FY-4A and FY-4B are geostationary and FY-3D, FY-3E and FY-3F are polar orbiting satellites.

FY4-A and 4-B carry three instruments:

AGRI - Advanced Geosynchronous Radiation Imager

GIIRS - Geostationary Interferometric Infrared Sounder for temperature and humidity profiles

LMI - Lightning Mapping Imager

China's polar orbiters FY-3D, FY-3E and FY-3F carry a wide variety of instruments:

ASI - Atmospheric Sounding Interferometer

ERM-2 - Improved Earth Radiation Measurement

GNOS - GNSS Radio Occultation Sounder

HIRAS - Hyperspectral Infrared Atmospheric Sounder

MERSI-LL - Medium Resolution Spectral Imager - LowLight

MWHS-2 - Improved MicroWave Humidity Sounder

MWTS-2 - MicroWave Temperature Sounder-2

WindRAD - Wind Radar

ASI, GNOS, HIRAS, MWHS-2 and MWTS-2 measure the atmospheric temperature and humidity profiles, WindRAD measures the sea surface winds, Ozone is measured by ASI and MWHS-2, and ERM-2, SIM-2, SSIM measure the radiation budget components. MERSI-LL measures land vegetation and ocean colour. Further details are available at https://database.eohandbook.com/database.

6.7 Russia – Electro and Meteor

Russia's geostationary Elektro-L N3 located over the Indian Ocean at 76 °E and Elektro-L N4 over the western Pacific at 165.8 °E carry basically three payloads, DCS, GGAK-E and MSU-GS. Of these, DCS is a data collection system and GGAK-E is a module for gravity, geomagnetic and geodynamic measurements.

MSU-GS which is a multispectral scanning imager-radiometer, is the only meteorological payload. It measures cloud type, amount and cloud top temperature, atmospheric temperature and humidity fields, winds, liquid water and precipitation rate, ozone, radiation budget parameters, sea and land surface temperature, snow and ice.

Russia's polar orbiting meteorological satellites Meteor-M N2-3 and Meteor-M N2-4 carry the following payloads:

DCS - Data collection system

GGAK-M - Module for geophysical measurements

IKFS-2 - IR-Fourier spectrometer

KMSS - Multispectral imager VIS

MSU-MR - Low-resolution multispectral scanning imager-radiometer

MTVZA-GY - Scanning microwave imager-sounder

Severyanin-M – Imaging radar

Barring DCS and GGAk-M, all the other instruments put together measure cloud type, amount and cloud top temperature, cloud particle properties, liquid water and precipitation rate, atmospheric temperature and humidity fields, ozone, radiation budget components, aerosols, trace gases, land albedo, vegetation, land surface temperature, SST, sea surface wind, snow and ice cover.

6.8 South Korea – Geo-KOMPSAT

S. Korea has one primary operational geostationary meteorological satellite named as GEO-KOMPSAT-2B (Korean Communications, Ocean and Meteorological Satellite) which covers the Asia-Pacific region, including the Korean Peninsula (Chung et al 2024).

It carries two instruments: the Advanced Meteorological Imager (AMI) and the Korean Space Environment Monitor (KSEM). It also carries two other instruments, the Geostationary Environment Monitoring Spectrometer (GEMS) and the Advanced Geostationary Ocean Colour Imager-II (GOCI-II).

AMI has 16 spectral channels between 0.47 and 13.3 μ, ranging from VIS to TIR with spatial resolution of 1 km for VIS and 2 km for IR. The scan takes 15 minutes for the full disc mode and 30 sec for the local area mode,

GEMS is a hyperspectral imager, with a spectral range of 300 - 500 nm and spectral resolution of less than 0.6 nm, sampling at less than 0.2 nm intervals. GOCI-II operates in 12 VIS and NIR bands, with one additional wideband, covering a spectral range of 380-900 nm. The instrument can scan an area of 2500 km^2 of the North-East Asian Seas in local area mode, with a spatial resolution of 300 m.

6.9 WMO Space Programme

Considering that the global meteorological satellite configuration is made up of a total of 27 satellites launched and maintained by 7 countries and agencies, the need to have coordination amongst them is obvious. In this regard, the World Meteorological Organization (WMO) plays a crucial role. The WMO Space Programme coordinates and promotes the utilization of satellite data and products to derive maximum benefits from them for all WMO application areas in weather, water, climate and environmental services, especially in support of WMO member countries having limited resources (Balogh and Toshiyuki 2020).

The WMO Space Programme works as a space-based observing system component of the WMO Integrated Global Observing System (WIGOS). One of its activities is to keep under review the allocations of radio-frequency bands to meteorological services for ground- and space-based observation systems. WMO partners with satellite operating agencies, through its Coordination Group for Meteorological Satellites (CGMS) and the Committee on Earth Observation Satellites (CEOS).

The WMO Space Programme also supports international coordination of operational activities to improve space weather hazard warnings for key application sectors such as aviation, radio communications, satellite operations, and electric power delivery.

6.10 Satellite Data Assimilation in Numerical Weather Prediction

The primary users of satellite imagery and products are naturally the meteorological agencies of the countries that have launched those satellites. They rightfully have immediate and full access to the data and they freely use them for purposes of forecasting, climatology and research. A lot of satellite imagery and information is shared in near real time with designated agencies and disseminated globally via the internet for public use. However, what truly makes a difference to the accuracy of weather forecasts is the contribution of satellite data to numerical weather prediction models

Notwithstanding the synoptic observations recorded and exchanged by national meteorological services under the auspices of WMO and its various programmes, satellite data are vital to the operational NWP models. This is because of their unmatched near-continuous global coverage, especially over the oceans, the polar regions and remote and inaccessible areas. When satellite data assimilation in NWP models first began, it was essentially to use the vertical temperature profiles and cloud motion winds to fill up the gaps in conventional data. Later on, however, it was realized that this did not necessarily improve the forecast skills and even had adverse effects, the reason being the errors and biases inherent to the satellite retrievals. This led to the new thinking that it might be better to assimilate the basic satellite radiances themselves rather than products derived indirectly from them (Eyre et al 2022). The main advantage of assimilating satellite radiances is that the errors in radiances are much easier to characterize than the errors in the retrievals. In the recent past, there has also been such remarkable progress both in satellite technology and in the product retrieval methods that 4D assimilation of both satellite radiances and products has now become indispensable to NWP. The four-dimensional variational (4D-Var) assimilation system adjusts the model state over a time window to best fit the observations.

Satellite data assimilation has become a crucial component of the forecasting system at the European Centre for Medium Range Weather Forecasts (ECMWF), the U.S. National Centers for Environmental Prediction (NCEP) and India's National Centre for Medium Range Weather Forecasting (NCMRWF). The satellite data that have most impacted NWP forecasts are the radiances of geostationary satellites such as GOES-R, MTG and Himawari, and products of various hyperspectral, microwave and RO sounders. ECMWF alone routinely processes data from around 90 satellite instruments.

ECMWF employs the technique of Ensemble of Data Assimilations (EDA) to estimate uncertainties, and runs 51 perturbed versions of the 4D-Var system. This ensemble helps quantify analysis errors and supports probabilistic forecasting. NCEP uses a 4D Hybrid Ensemble-Variational system. NCMRWF operates two major global data assimilation systems: Next Generation Forecast System – Global Data Assimilation System (NGFS-GDAS) and NCMRWF Unified Model (NCUM).

Eyre et al (2022) have reviewed the developments in the assimilation of satellite data in NWP from the 1960s to the 2020s. The progress is attributable to advances in the satellite remote sensing technologies, in the sophistication of radiative transfer models, and in more effective methods to assimilate satellite data into NWP systems.

6.11 References

Balogh W. and Toshiyuki K., 2020, "The World Meteorological Organization and space-based observations for weather, climate, water and related environmental services", *Space capacity building in the XXI century*. Springer International Publishing, 223-232.

Chung Sung-Rae et al, 2024, "History, status, and prospects of remote sensing in the field of meteorological satellite in Korea", *Korean J. Remote Sensing*, 40, 713-726.

Cohen M. et al, 2006, "The Eumetsat Polar System – a major step for operational meteorology", *ESA Bull.*, 127, 18-23.

Eyre J.R., et al, 2022, "Assimilation of satellite data in numerical weather prediction. Part II: recent years." *Quart. J. Royal Meteor. Soc.,* 148, 521-556.

Lee T. E. et al, 2006, "The NPOESS VIIRS day/night visible sensor", *Bull. Amer. Meteor. Soc.,* 87, 191-199.

Lee T. F. et al, 2010, "NPOESS: Next-generation operational global earth observations", *Bull. Amer. Meteor. Soc.*, 91(6), 727-740.

Miller S. D. et al, 2006, "NexSat - previewing NPOESS/VIIRS imagery capabilities", *Bull. Amer. Meteor. Soc.*, 87, 433-446.

Schmetz J. et al, 2012, "Meteosat Third Generation (MTG) development in the context of other future geostationary satellite observations", *AGU Fall Meeting Abstracts,* 1, 4.

Schmit T. J. et al, 2005, "Introducing the next-generation advanced baseline imager on GOES-R", *Bull. Amer. Meteor. Soc.*, 86, 1079-1096.

Schmit T. J. et al, 2017, "A closer look at the ABI on the GOES-R series", *Bull. Amer. Meteor. Soc.*, 98, 681-698

Chapter 7

Satellite Monitoring of Climate Change

Historically speaking, the first beneficiaries of satellite imagery and products were synoptic meteorologists who considered them as value additions to their weather charts, especially over data sparse regions and oceans. It soon became apparent, however, that the utility of satellites was not confined to gathering temperature, humidity and wind data for purposes of operational weather forecasting. A different category of satellites soon emerged that would gather data needed by climate researchers. Many missions with experimental payloads continued long beyond their expected lives giving an insight into the processes that cause climate change. In recent years, several satellites have been launched for the measurement of cloud properties, radiative fluxes, greenhouse gas concentrations, aerosols, sea level rise, ice and snow cover, and land use. They monitor these parameters on a global scale and their data is being used as an input to climate models.

7.1 Earth Radiation Budget

The energy that drives the entire climate system of the earth comes from a single source, the sun. The annual average solar irradiance received outside the earth's atmosphere on a surface of unit area normal to the incident radiation and at the mean sun-earth distance has been defined as the solar constant. There is a global network of surface observations at which the direct and diffuse components of solar radiation as received at the ground are regularly measured. Obviously, it is not possible to make similar measurements at the top of the atmosphere. In the absence of data, it has always been a matter of speculation whether the solar constant is indeed a constant and whether variations in the solar constant can cause the earth's climate to change. Ground-based measurements of the solar constant date back to 1913 but the difficulty with such measurements is that they must be corrected for the effect of the intervening atmosphere. Only satellite-based measurements are capable of providing more accurate values of the solar constant since they are made outside the earth's atmosphere.

The Nimbus-6 and later satellites in the series carried an instrument. that could view the sun for approximately 3 min and measured the solar irradiance in several spectral channels. The Solar Maximum Mission satellite, launched in 1980 carried an Active Cavity Radiometer Irradiance Monitor (ACRIM).

The mean Nimbus-7 solar constant, measured over an 11-year solar cycle, was 1372 W m^{-2} with a standard deviation of 0.6 W m^{-2}. The Solar Maximum Mission gave a mean solar constant over a period shorter than a solar cycle of 1367.5 W m^{-2} with a standard deviation of 0.4 W m^{-2}. The value obtained from ERBE measurements is approximately 1365 W m^{-2}. Such differences of the order of 7 W m^{-2} across various satellites are only to be expected because of the difficulty in achieving an absolute calibration. However, the measurements made by the different instruments are highly correlated and the relative variations in the solar constant are in good agreement (Lean et al 1998).

The ACRIMSAT satellite was launched in December 1999 and it carried an advanced version of the Active Cavity Radiometer Irradiance Monitor (ACRIM-III). This satellite provided the first data set that clearly demonstrated that the total radiant energy from the sun was not a constant.

The Solar Radiation and Climate Experiment (SORCE) was a NASA-sponsored satellite mission for making highly accurate measurements of incoming solar radiation separately in the x-ray, ultraviolet, visible, near-infrared, and total wavelength bands. SORCE was launched in January 2003 into a 645 km, 40° inclination orbit. It carried four instruments, Total Irradiance Monitor (TIM), Solar Stellar Irradiance Comparison Experiment (SOLSTICE) covering 0.1-0.3 μ, Spectral Irradiance Monitor (SIM) covering 0.2-2 μ and X-Ray and UV Photometer System (XPS). With the sophisticated measurements provided by SORCE in different spectral bands it was possible to make specific studies of how variations in solar radiation could possibly influence climate change and the data could be incorporated in climate prediction models.

7.1.1 ERBE, CERES and GERB

The Earth Radiation Budget Satellite (ERBS), launched in 1984 aboard the Space Shuttle Challenger, was one of NASA's earliest missions designed to study how the energy from the sun is absorbed and re-emitted by the earth and its atmosphere. This information is crucial to our understanding of climate and climate change. The ERBS payload consisted of ERBE (Earth Radiation Budget Experiment) Scanner and Non-Scanner Radiometers to measure shortwave and longwave radiation, and SAGE II (Stratospheric Aerosol and Gas Experiment) to measure aerosols and gases in the stratosphere. These measurements were of timely help in the negotiations that led to the adoption of the Montreal Protocol on ozone depletion. The ERBS payloads were later replicated on the NOAA-9 and -10 satellites as well, thus making a long continuous data set. A new instrument named CERES (Clouds and the Earth's Radiant Energy System) was flown on several satellites subsequently such as Terra, Aqua, Suomi NPP, NOAA-20 and NOAA-21.

The CERES instrument operates in three channels, SW (0.3–5 μ): for reflected sunlight, TOT (0.3–200 μ) for total radiation and WIN (8–12 μ) or LW (5–35 μm): for thermal emission. The field of view is 10–24 km depending on the satellite. There are three scanning modes, cross-track, rotating azimuth, and programmable azimuth.

ERBE's geostationary counterpart is GERB (Geostationary Earth Radiation Budget) which is flown on Meteosat second and third generation satellites. This is a broadband visible-infrared radiometer having a spectral coverage of the solar spectrum 0.32 to 4.0 μ and the thermal spectrum 0.32 to 30 μ.

7.2 Atmospheric Ozone

The ozone layer that resides in the earth's stratosphere acts as a natural shield that protects living beings on earth from the harmful effects of the ultraviolet radiation of the sun. As any depletion of stratospheric ozone will render us more vulnerable, ozone concentrations need to be monitored carefully on a global scale. There is an extensive network of spectrophotometer stations that make ground-based measurements of total column ozone and surface ozone levels. At some locations, vertical ozone profiles are being obtained with ozonesondes. As with most other networks, the availability of data is confined to land regions.

7.2.1 TOMS, OMI and GOME

The earliest satellite measurements of atmospheric ozone were made between 1978 and 1994 by the Total Ozone Mapping Spectrometer (TOMS) on the U.S. satellite Nimbus-7 and the Russian satellite Meteor-3. Another TOMS instrument was launched in July 1996 on the NASA Earth Probe. The total ozone amount was computed by comparing the incoming solar energy and the backscattered UV radiation measured by TOMS at six different wavelengths. Another TOMS instrument was launched on the Japanese ADEOS satellite in September 1996 but the mission was short-lived. The Ozone Monitoring Instrument (OMI) on board NASA's Aura satellite has been in operation since 2004. OMI employed hyperspectral imaging in a push-broom mode, thus improving upon the accuracy and precision of TOMS measurements.

In April 1995, the European Space Agency (ESA) launched the Global Ozone Monitoring Experiment (GOME) on the second European Remote Sensing satellite ERS-2. GOME had four spectral channels 237-316, 311-405, 405-611 and 595-793 nm in the UV, VIS and NIR wavelength regions respectively, with a spectral resolution of 0.20-0.33 nm.

GOME was followed by a more comprehensive mission known as SCIAMACHY, launched on the ENVISAT-1 platform in 2002. SCIAMACHY was an acronym for Scanning Imaging Absorption Spectrometer for Atmospheric CHartographY. This was a spectrometer designed to measure solar radiation and its reflection and scattering by the earth's surface and atmosphere, in eight channels in the UV, VIS and NIR wavelength regions between 240 and 2380 nm at a moderate spectral resolution of 0.2 -1.5 nm. Both the GOME and SCIAMACHY missions have come to an end. Currently a new GOME-2 instrument is flying on the MetOp satellite and an Ozone Monitoring and Profiler Suite (MPS) on the Suomi NPP satellite.

The sounders on the NOAA polar orbiting satellites and INSAT-3D geostationary satellites which have a 9.7 μ channel that peaks at the wavelength of maximum absorption by stratospheric ozone, have also been providing global total ozone measurements (Kumar et al 2021, Sangeetha et al 2025).

7.2.2 Antarctic Ozone Hole

Satellite measurements of stratospheric ozone have greatly helped to generate detailed maps of the global ozone distribution and resulted in the discovery of what has come to be popularly known as the Antarctic ozone hole. The ozone hole is actually an area of exceptionally low concentrations of stratospheric ozone, which begins to develop in August and continues to spread in September and October.

The annual formation of the ozone hole is linked to the polar vortex that develops over the southern hemisphere in austral winter. The polar vortex is associated with strong winds, extremely cold temperatures and polar stratospheric clouds. As spring returns, ozone depleting substances react to sunlight on the surface of these clouds releasing chlorine and bromine atoms that break ozone molecules down into oxygen molecules and atoms.

The stronger the polar vortex, the larger and deeper is the ozone hole, and vice versa. With the arrival of austral summer and increased sunlight the stratosphere warms up again, disrupting the polar vortex, and ultimately closing the ozone hole as mid-latitude air with higher ozone concentration is able to reach the polar latitudes. Global warming and the greenhouse effect tend to cool the stratosphere favouring the formation of polar stratospheric clouds and ozone depletion.

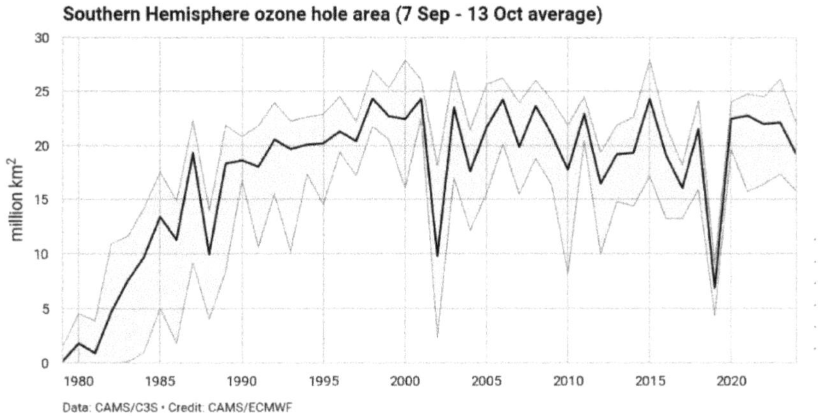

Figure 7.2.2.1 Variations in the ozone hole area from 1979 to 2024. (Source: CAMS/ECMWF)

In 1987, the Montreal Protocol was adopted, mandating the phase-out of ozone-depleting substances like CFCs. While seasonal ozone holes still occur, signs of recovery have emerged thanks to reduced CFC emissions. Figure 7.2.2.1 shows the time series of the average ozone hole area from 1979 to 2024. The smallest ozone holes were observed in 1988, 2002 and 2019. The ozone layer is seen to be on the path of recovery, with the situation expected to revert back to the 1979 level by the year 2066.

7.3 Sea Level Rise

The sea surface, or the boundary between the atmosphere and the ocean, is continuously changing because of winds, tides and waves. When such fluctuations are averaged out over a long period of time, we get the mean sea level, which is commonly used as a reference point for expressing the height of a place, especially a mountain. Until recently, mean sea level was of interest mainly in geology and geography, but sea level rise is now regarded as a result of global warming, and therefore being closely monitored. The reasons behind are the thermal expansion of the ocean and the increasing inflow of water into the ocean arising from melting of glaciers. Sea level rise will not only have consequences for the ocean-atmosphere system itself, but it also poses a serious threat to the populations residing in coastal regions and islands. It will lead to coastal erosion, intrusion of salt water, loss of mangroves and increased vulnerability to storms and flooding.

Historical sea level observations were made with tide gauges installed at coastal stations. The global tide gauge network was sparse and the records were vitiated by relative land movements. Satellite measurements of sea surface height first began in 1992 with the TOPEX/Poseidon satellite which was a joint U/S.-French venture (Fu et al 1994). Many countries have since collaborated to not only keep these measurements going but to improve their accuracy and coverage with satellites like Jason-1 launched in 2001, Jason-2 in 2008, Jason-3 in 2016 and Sentinel-6 Michael Freilich in 2020 (Le Traon et al 2025). The last two are now operational and Sentinel-6B is likely to go into orbit by 2025 end. Incidentally, Sentinel-6 is named in honour of Michael Freilich, who was a renowned oceanographer and headed NASA's Earth Science Division from 2006 to 2019.

For measuring sea surface height, satellites use an altimeter, which is a radar operating at a microwave frequency in which the sea surface acts as a good reflector. The radar pulse illuminates a spot on the sea surface which is large enough to smooth out the effects of local waves and ripples, and a high pulse repetition rate is used to improve the signal-to-noise ratio, especially when the ocean surface is rough. The time taken by the pulse to return to the satellite is a measure of the sea surface height.

The aim here is to detect changes in the sea level of the order of millimetres from a satellite altitude of several hundred kilometres. The process demands extreme precision and the satellite position itself has to be tracked most accurately. Corrections have to be applied to the travel time of the pulse to account for ionospheric and atmospheric delays, and known tidal corrections are made as well. The sea surface height is calculated as the difference between the distance of the satellite from the centre of the earth and the distance between the sea surface and the satellite as measured by the altimeter.

Sentinel-6 Michael Freilich is flying in a non-sun-synchronous low earth orbit at an altitude of 1336 km. The orbit inclination is 66° providing wide coverage of the earth's oceans with a repeat cycle of 10 days. Its principal payload is the Poseidon-4 dual frequency (C and Ku-band) nadir-pointing radar altimeter. The C-band frequency is 5.41 GHz and the Ku-band frequency is 13.575 GHz. It has another 3-channel Advanced Microwave Radiometer for Climate (AMRC) which includes a High Resolution Microwave Radiometer (HRMR) operating in millimetre-wave channels at 90, 130 and 168 GHz. This provides measurements of atmospheric water vapour to mitigate the degradation of the radar altimeter measurements. There is also a Precise Orbit Determination (GNSS-POD) Receiver. A complete description of the satellite and its payloads has been given by Donion et al (2021).

Hamlington et al (2024) have investigated the continuous, accurate and near-global scales measurements of sea surface height that satellites have made over the last three decades. From 1993 to 2023, global mean sea level has risen by 111 mm (Figure 7.3.1). Further, the rate of global mean sea level rise over this period has increased from 2.1 mm/year in 1993 to 4.5 mm/year in 2023. If this trajectory of sea level rise continues over the next three decades, sea levels are likely to increase by an additional 169 mm globally.

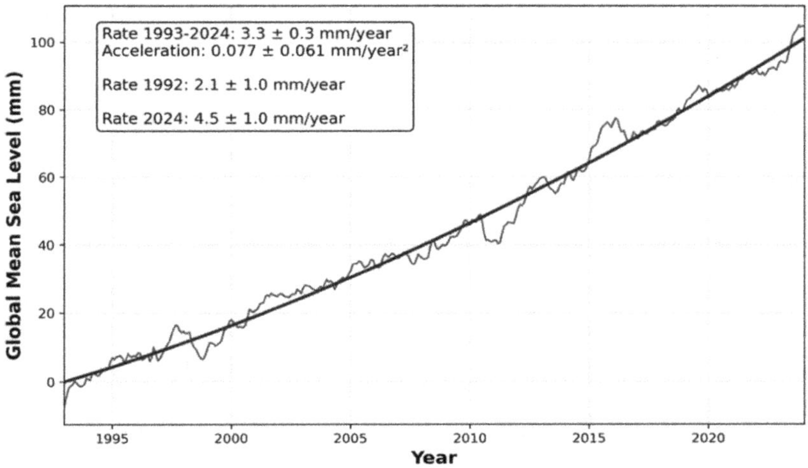

Figure 7.3.1 Satellite-derived global mean sea level rise in mm with 1993 as base year. (Source: Hamilton et al 2024)

In February 2013, ISRO had launched the SARAL satellite (Satellite with ARgos and ALtiKa), a joint India-France collaborative mission. This carried an altimeter called AltiKa, which worked in Ka-band at 35.75 GHz. AltiKa had been developed by the French National Space Agency, CNES (Steunou et al 2015). It was the first space-borne altimeter to operate in Ka-band. Its primary application was in oceanographic studies but it was also to be used for ice sheet and inland waters monitoring. The advantage of AltiKa's high frequency was that it had become more compact in size and unlike other altimeters, it did not need another instrument to correct for atmospheric effects. Operating at Ka-band frequencies also enabled AltiKa to measure the sea surface height with an accuracy as high as 8 mm which was far greater than that of other altimeters. The disadvantage of Ka-band, however, was that the measurements were extremely sensitive to rain, as the radar signal was attenuated by atmospheric liquid water. Cloud and rain effects might have vitiated a small part of the data which needed to be corrected. Although the mission life was expected to be five years, the spacecraft encountered a reaction wheel issue in 2016 and went into a drifting mode.

7.4 Sea Surface Temperature

The oceans of the world together occupy about 70 per cent of the surface area of the globe and play a major role in determining the earth's weather and climate. However, compared to the atmosphere, the ocean has been very poorly observed. It is only with the advent of satellite remote sensing that we have been able to substantially increase our knowledge of the ocean and to monitor it extensively and continuously.

Oceans and climate have a connection through complex exchanges of heat and mass. Oceans are the dominant heat reservoirs of the complete land-atmosphere-ocean system and play a critical role in climate and its variability. Ocean currents transport heat from the tropics to higher latitudes. Poleward heat transport by the oceans is of comparable magnitude with that of the atmosphere and shows large inter-annual variability.

Graphs depicting the steady rise of temperature beginning with the pre-industrial era to the present rely mostly on instrument readings at the surface either on land or ships. In the context of global warming, retrieval of sea surface temperature (SST) from satellite measured radiances is easier and more straightforward than land temperature and air temperature. Retrieved values of SST were prone to several types of errors when radiometers had a single IR channel. However, as time went by, most satellite radiometers had several IR channels and multi-channel SST retrievals became increasingly accurate.

Embury et al (2024) have meticulously compiled a 42-year climate data record of global satellite-derived SST covering the period 1980 to 2021 with a high degree of independence from in situ measurements. They have integrated continuous data series of many as 22 IR and 2 microwave radiometers. These are NOAA-AVHRR, NOAA-ATSR, MetOp-AVHRR and Sentinel-SLSTR. They have adjusted for their differing times of day of measurement. Data were processed at their native resolution, and averaged on a global latitude-longitude grid. Figure 7.4.1 shows the global SST variation over the 42-year period 1980 to 2021. There is a consistent warming trend with an overall increase of about 0.6°C.

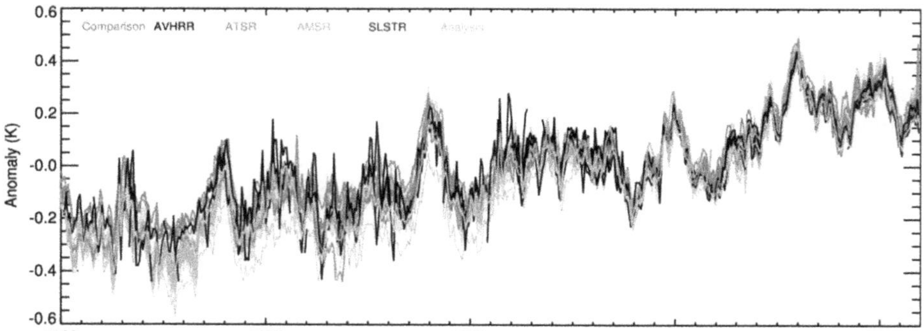

Figure 7.4.1 SST anomalies from 1980 to 2024. (Source: Embury et al 2024)

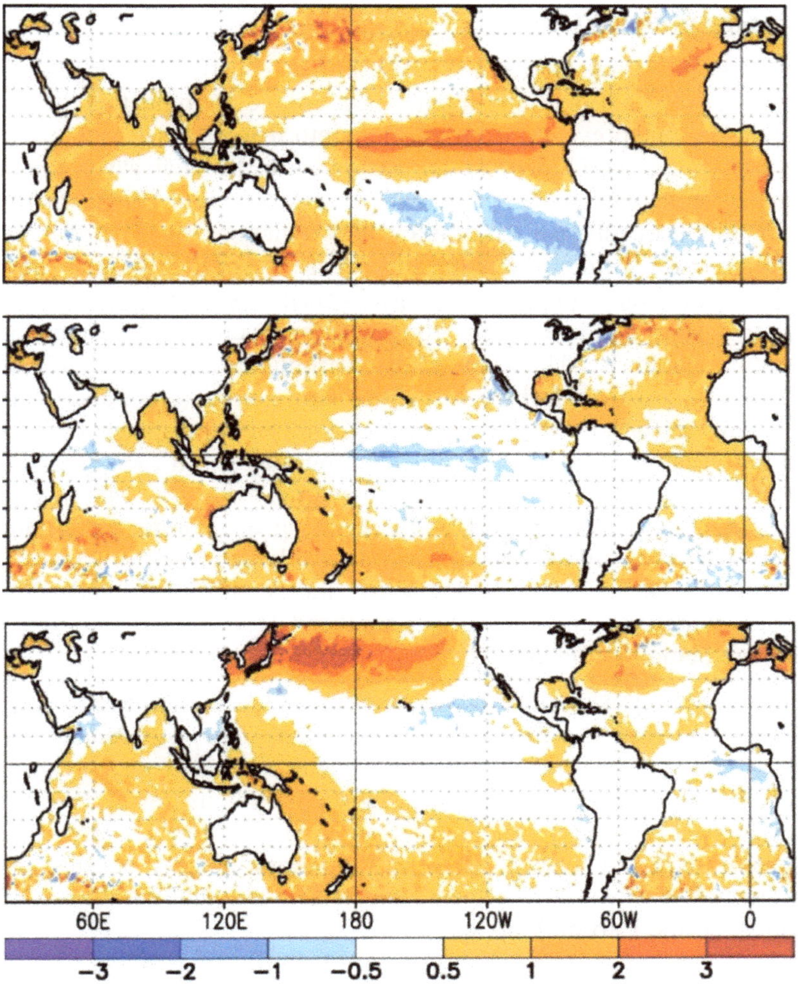

Figure 7.4.2 Global monthly SST anomalies in °C for
(top) December 2023 – El Nino
(centre) December 2024 – La Nina
(bottom) July 2025 – Neutral conditions.
(Source: https://www.cpc.ncep.noaa.gov)

SST anomalies are very useful to detect, monitor and predict phenomena like the El Nino and La Nina and the Indian Ocean Dipole which are important for the Indian monsoon. Figure 7.4.2 shows three global SST maps for the months of December 2023, December 2024 and July 2025, representative of El Nino, La Nina and neutral conditions.

7.5 Greenhouse Gases

Greenhouse gases or GHGs are very minor constituents of the earth's atmosphere but are of great importance because they trap heat emitted by the earth's surface and prevent it from escaping into space. This greenhouse effect is a natural and essential process for life on earth. In its absence, the earth's average temperature could go down to as low as -18°C, making it uninhabitable. Greenhouse gases allow the radiation from the sun to pass through the earth's atmosphere and then heat the earth's surface. They then absorb the radiation emitted by the earth's surface keeping the planet warm enough to sustain life. If the GHG concentrations were to increase the earth's temperature would rise. The recent global warming has been attributed mainly to two GHGs in the atmosphere, largely carbon dioxide (CO_2) and to a lesser extent methane (CH_4). Over the years, CO_2 in the atmosphere has kept on increasing because of human activities like burning fossil fuels such as coal, oil and natural gas, deforestation, and industrial processes. Methane comes from sources like livestock, landfills, rice farming, and natural gas production. Another GHG is nitrous oxide (N_2O) that is released mainly from fertilizer application in agriculture, besides fossil fuel combustion, and industrial processes. Water vapour (H_2O) is yet another GHG, its concentration in the atmosphere being highly variable.

While the monitoring of GHG concentrations has always been considered important in investigations of global warming, systematic ground measurements of GHGs were for several decades limited to only one gas, CO_2, and that too at only one location on earth, Mauna Loa in Hawaii (Figure 7.5). The monitoring of atmospheric GHGs from space was first attempted in 2009 but it has progressed further only in recent years, notably because of non-governmental enterprises that led to the launch of satellites like GHGSat, MethaneSAT and Carbon Mapper.

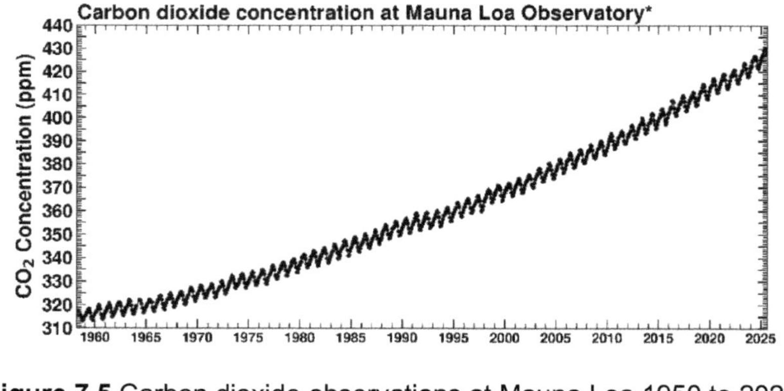

Figure 7.5 Carbon dioxide observations at Mauna Loa 1950 to 2025. (Source: The Keeling Curve web site https://keelingcurve.ucsd.edu/)

7.5.1 OCO

NASA launched the Orbiting Carbon Observatory (OCO) in February 2009 with a payload of three high-resolution grating spectrometers to measure CO_2 from space. The launch was, however, a failure. OCO-2 was launched in July 2014 and its data has been used extensively to demarcate the CO_2 sources and sinks around the globe, and to understand how emissions are offset by natural carbon sinks like forests and oceans. OCO-3 is a similar NASA instrument which was mounted on the International Space Station (ISS) in May 2019.

Both OCO-2 and -3 use three high-resolution grating spectrometers to measure CO_2 concentrations with a precision of 1 ppm. The orbit of OCO-2 is different from that of the ISS, so their observation times and coverage are different. The spectrometer system analyzes the sunlight reflected from the earth to detect CO_2 and oxygen absorption at different wavelengths. The Solar-Induced Fluorescence instrument detects the faint light emitted by chlorophyll during photosynthesis, helping to assess plant activity and carbon uptake. The OCO-3 on the ISS uses a snapshot mode with a 2-axis pointing mirror to scan large areas (up to 80 × 80 km) in minutes, which is ideal for targeting cities and dynamic regions.

7.5.2 GHGSat

GHGSat is a pioneering space mission, run by a commercial venture, that aims at monitoring greenhouse gas emissions, particularly carbon dioxide and methane from sources as small as individual oil wells and landfills, with high resolution. The GHGSat constellation currently consists of 12 satellites. The first of these was launched in June 2016 as a technology demonstrator satellite, and the other 11 satellites were launched between 2020 and 2023. Four more satellites are likely to be added in the future. Each GHGSat is equipped with a Wide Angle Fabry-Perot (WAF-P) Imaging Spectrometer designed to measure the vertical column abundances of GHGs and is capable of measuring both methane and carbon dioxide, but the present constellation is optimised for methane measurements (Ligori et al 2019).

The WAF-P Imaging Spectrometer is the primary sensor which produces a hypercube consisting of a stack of overlapping, spectrally selected images acquired within 30 seconds. This hypercube embeds, for each ground pixel, information equivalent to hundreds of wavelengths from the top-of-the-atmosphere spectral radiance. Once downloaded, the hypercube is corrected for sensor response and instrument optics. The gas column density and surface reflectance information are then retrieved for each ground pixel using a measurement model which includes surface, instrument and atmospheric contributions to radiance. The resulting arrays are geo-referenced, giving the surface reflectance image and an abundance dataset for methane.

The GHGSat satellites have a spatial resolution of 25 m and a field of view of 12 x 12 km, allowing gas plumes emitted by industrial sources to be captured and distinguished from the surrounding background concentrations, constituting a differential measurement. The reflectance and abundance products can be combined into a high readability concentration map for human interpretation. The abundance datasets from multiple observations of a given site can be combined to form a monitoring product for semi-automatic detection of emitter activity or large changes in emission rates using pre-determined thresholds. Abundance datasets also allow emission rates to be estimated for individual sources. Plume dispersion is modelled using site information and weather data from global meteorological models or local weather stations and matched to the measured datasets (Ramier et al 2023).

7.5.3 MethaneSAT

Methane concentration in the earth's atmosphere is far less than that of carbon dioxide but it is 25 times more efficient as a greenhouse gas. Methane emissions are therefore a prime target in the effort to minimize the risks of global warming. Keeping this in view, MethaneSAT was launched on a SpaceX rocket in March 2024 by a non-profit organization, Environmental Defense Fund (EDF), to monitor methane leaks from oil and gas operations, and then making the data available to policymakers and scientists through open access. The spacecraft was placed in a sun-synchronous orbit at an altitude of 590 km, with a 95 min orbital period and a swath width of 200 km.

MethaneSAT carried two Passive Infrared Littrow Spectrometers, one covering 1249-1305 nm wavelength range for oxygen detection, and another covering 1605-1683 nm wavelength range for methane and carbon dioxide retrievals, with 0.1 nm spectral sampling and 0.3 nm spectral resolution. The spectrometers were capable of detecting changes in gas concentrations in the atmosphere as small as 3 parts per billion with a native pixel size of 100 x 400 m (Rohrschneider et al 2021).

Unfortunately, after a year of successful operation, contact with the satellite was lost in June 2025 and attempts to recover it had failed.

7.5.4 Tanager-1

Carbon Mapper is a non-profit organization whose mission is to help reduce GHG emissions by making methane and carbon dioxide data accessible and actionable. With this aim, it has planned a new constellation of hyperspectral satellites, the first of which, named Tanager-1, was launched in August 2024. The constellation is likely to be expanded to comprise four Tanager satellites.

The system is designed to balance detection limits, spatial coverage and sample frequency to optimize observational completeness for high emission methane and carbon dioxide point sources. Each imaging spectrometer instrument has a spectral range of about 400–2500 nm, 5 nm spectral sampling, a nadir spatial resolution of 30 m, and nadir swath width of about 19 km. The Carbon Mapper monitoring strategy focuses on routine mapping of major methane and carbon dioxide emitting regions around the globe. The access time for a single Tanager satellite ranges from 2 to 5 days. With all four Tanager satellites, the cloud-free median access time will decrease to 10 hours. The satellites are also capable of sun glint tracking over the ocean to allow monitoring of offshore oil and gas fields, a key emission sector that has largely gone unmonitored.

A review of the first seven months of Tanager-1 methane and carbon dioxide observations including initial validation with coordinated aircraft under-flights has indicated that the system is meeting performance requirements (Duren et al 2025).

7.6 Aerosols

In recent years, an increasing importance is being given to the study and measurement of aerosols in the atmosphere, because of their role in climate processes which is considered crucial but not yet completely understood. Aerosols are produced as a result of biomass burning or combustion of fossil fuels, and industrial activity and vehicle emissions are their main sources. There are also natural aerosol sources like breaking sea waves or wind-blown dust. Absorption of radiation by aerosols like silicates, sulphates and nitrates, results in a heating of the atmosphere while scattering of radiation by aerosols leads to cooling. The presence of aerosols in the atmosphere therefore not only pollutes the environment locally but also alters the radiative balance and it can directly influence the global climate. Aerosols can also have an indirect effect on climate processes as they modify cloud properties like droplet size. However, our present state of knowledge of the properties of different aerosols and their distribution over land and oceans is full of uncertainties which need to be reduced if the results of climate prediction models are to be made more accurate.

Satellite monitoring of aerosols from space involves conversion of the satellite-measured radiance into different aerosol properties. However, this is not a straightforward process as the same value of the radiance may be arrived at through different combinations of aerosol properties. Moreover, radiation reaching the satellite would have undergone not only absorption and scattering but also a change in its polarization. Although aerosols are also present in the stratosphere, their concentration there is relatively much smaller than in the troposphere except after occurrences of volcanic eruptions.

The aerosol properties required to be estimated from satellite measurements are the aerosol optical depth (AOD), phase function and single scattering albedo, all as a function of wavelength, the aerosol columnar concentration, aerosol size distribution and chemical composition. While available satellite sensors like NOAA AVHRR and MODIS are being used for this purpose, POLDER was one instrument which was specially designed for aerosol measurements (see Section 4.2.1). Instruments like Stratospheric Aerosol and Gas Experiment (SAGE) were used earlier for monitoring stratospheric aerosols.

7.7 References

Donion C. J. et al, 2021, "The Copernicus Sentinel-6 mission: enhanced continuity of satellite sea level measurements from space", *Remote Sensing of Environment*, 258, 112395.

Duren R. et al, 2025, "The Carbon Mapper emissions monitoring system", *EGUsphere*, https://doi.org/10.5194/egusphere-2025-2275.

Embury O., et al, 2024, "Satellite-based time-series of sea surface temperature since 1980 for climate applications", *Sci. Data,* 11, 326.

Fu L.L. et al, 1994, "TOPEX/POSEIDON mission overview", *J. Geophys. Res. Oceans,* 99, 24369-24381.

Hamlington B. D. et al, 2024, "The rate of global sea level rise doubled during the past three decades", *Communications Earth Environment*, 5, 601.

Kumar R.R. et al, 2021, "Comparison of INSAT-3D retrieved total column ozone with ground-based and AIRS observations over India", *Science of the Total Environment*, 793, 148518.

Lean J. and Rind D., 1998, "Climate forcing by changing solar radiation", *J. Climate,* 11, 3069-3094.

Le Traon et al, 2025, "Satellite altimetry and operational oceanography: from Jason-1 to SWOT", *Ocean Sci.*, 21, 1329-1347.

Ligori M., Bradbury L., Spina R. and Zee R.E., 2019, "GHGSat Constellation: The Future of Monitoring Greenhouse Gas Emissions", *33rd Annual AIAA/USU Conf Small Satellites.*

Ramier A. et al, 2023, "High-resolution methane detection with the GHGSat constellation", IAF Global Space Conf Climate Change, Oslo.

Rohrschneider, R. R., et al. 2021, "The MethaneSAT Mission", Conf Small Satellites.

Sangeetha P., Gopikrishnan G.S., Kuttippurath J. and Thapliyal P. K., 2025, "Evaluation of total column ozone measurements from INSAT-3D/3DR with satellite-based AIRS and CAMS reanalysis data", *J. Indian Society of Remote Sensing*, 1-12.

Steunou N., et al, 2015, "Altika radiometer: instrument and in–flight performance", *Marine Geodesy*, 38.sup1, 43-61.

Chapter 8

New Dimensions in Satellite Remote Sensing

The many stages of evolution that satellite meteorology has gone through its not-so-long history are evident from the previous chapters of this book. The first pictures from space of the earth and its cloud cover were taken by astronauts with their cameras through the windows of their space capsules. Then came dedicated weather satellites with television cameras that relayed to the ground the scenes that they viewed. These were soon replaced by scanning radiometers that measured visible and radiances that could be digitally converted into the more familiar image form. As technology advanced, more spectral channels were added and a significant transition was to the use of hyperspectral imaging. Passive microwave sensors have given way to synthetic aperture radars. In parallel, satellite orbits have improved progressively from random over-passes to sun-synchronized visits, settled down for a long time for a combination of polar orbiting and geostationary orbits, until technology paved the way for placing satellites in tropical and subtropical orbits at low elevations.

The rapid growth of satellite meteorology is the result of a highly complex interplay of scientific thinking, technological innovation and optimum use of resources. Many a times, revolutionary concepts were proposed for consideration by the scientific community and even found acceptance, but they just had to wait. Some of them even went into permanent cold storage. On the other hand, it also happened that a small group of scientists and engineers working in a little known institution were able to get their payload on a spacecraft and their mission became very successful.

The greatest need of the present time is to improve our understanding of the world's climate, to detect, attribute and understand climate change, to reduce uncertainties, and to project future climate change. The trend now is towards integrated remote sensing that would provide accurate, long-term and consistent data. Changes in radiative forcing, spatial distribution of greenhouse gases, aerosols and CO_2 sources and sinks, and measurement of precipitation, clouds, surface exchanges, and interactions of these with radiation can now be monitored together by a single satellite. The high degree of accuracy and resolution with which these parameters are now measured was previously unthinkable.

On the other hand, the concept of small satellites has also come into vogue and there is a radical change in the way these satellites are made, launched, and operated. They cost so little that even students can fabricate them, They can be launched with small rockets or along with large spacecrafts. Private enterprises can afford to launch them in large numbers in a constellation mode, taking risks of some degree of failure.

8.1 Small Satellites

Small satellites can be of the size of a shoebox and may weigh anywhere between 1 and 100 kilogram depending upon the nature of the payload. They are useful for testing a new instrument which may be risky to launch on a major mission. The power of small satellites is truly manifested when they are launched in large numbers as a constellation. The PlanetScope network has presently over 400 small sun-synchronous satellites in low altitude polar orbit that provide daily earth imagery at a resolution of 3.7 m. The Starlink (SpaceX) constellation has several thousands of low earth orbit satellites with inclinations between 53 and 98° that deliver broadband internet to remote regions of the world. Both of these are private commercial enterprises.

The concept of a small satellite was first mooted by NASA at the turn of the millennium and successfully demonstrated by it in October 2006 with the name CubeSat. NASA has so far launched or supported hundreds of CubeSat missions for a variety of purposes, including earth observation, lunar exploration, space weather monitoring, technology demonstrations, and scientific research.

RainCube was NASA's first CubeSat to carry a meteorological payload, which was a Ka-band 5.75 GHz precipitation profiling radar designed to study storms (Peral et al 2015), RainCube was deployed in July 2018 from the International Space Station and remained in orbit until December 2020.

The Cyclone Global Navigation Satellite System (CYGNSS) is another NASA mission launched in December 2016 with a constellation of eight microsatellites aimed at measuring ocean surface wind speeds, especially in the inner core of tropical cyclones. It uses the technique of passive GNSS reflectometry to measure reflected GPS signals from the ocean surface. CYGNSS satellites are in orbits inclined at 35° so they can cover regions of the earth which are most prone to tropical cyclones with revisit times as short as 3–7 hours (Carreno-Luengo et al 2021).

INS-1A and INS-1B were two experimental nanosatellites developed by ISRO as part of its Indian NanoSatellite (INS) program. Weighing only about 9 kg, they were launched together aboard a PSLV in February 2017, basically as demonstration missions. INS-1A carried a radiometer that measured sunlight reflected by the earth's surface.

8.1.1 Precipitation Structure of Tropical Cyclones

What a satellite can deliver in terms of useful data depends upon its orbit, coverage, repetivity, and the resolution and spectral characteristics of the instrument it carries. It is therefore impossible to have a single satellite that will serve all applications. The design and development of meteorological satellites has always been a matter of compromise. For example, geostationary satellites help us to detect tropical cyclones far out at sea, make a coarse estimate of their intensity and track their movements round the clock. Satellites such as CloudSat could show a vertical cross-section through a tropical cyclone but they would pass over the cyclone only twice a day. NASA's TROPICS mission is an attempt to achieve both, using a network of small satellites.

8.1.2 TROPICS

The acronym TROPICS stands for Time-Resolved Observations of Precipitation structure and storm Intensity with a Constellation of Smallsats. Its objective is to make frequent measurements in a tropical cyclone situation through the overall cloud cover to see the storm's underlying structure, and predict its likely intensification (Blackwell et al 2018). It is one of NASA's science-driven, low-cost earth venture missions. TROPICS was originally conceived as a constellation of six small satellites. However, the first two satellites were lost during a launch failure in June 2022. Thereafter, without compromising its objectives, the mission was redesigned to work with four satellites and all of them were launched in May 2023. The TROPICS CubeSats are incredibly small, each measuring only 36 ×10 × 10 cm, and weighing barely 5 kg. The TROPICS satellites are neither geostationary nor polar orbiting. They are placed in 30° inclination low earth orbits at 550 km altitude to cover cyclone-prone latitudes of the earth. They are distributed two each in two orbital planes, to enable scanning a tropical cyclone as frequently as every 20 minutes.

Each of the TROPICS satellite has as its payload a 12-channel scanning passive microwave radiometer. The 12 channels are so chosen as to help retrieve temperature and humidity profiles, precipitation and cloud properties in the tropical cyclone environment. There are 7 channels near the 118.75 GHz oxygen absorption line for temperature profiles and 3 channels near the 183 GHz water vapour absorption line for water vapour profiles. There is a 90 GHz channel for precipitation measurements and a 206 GHz channel for cloud ice measurements.

This observing system offers an unprecedented combination of spatial and temporal resolution to measure environmental and inner-core conditions in and around tropical cyclones in whichever basin they may form. The TROPICS data is expected to result in a significant improvement in our understanding of the precipitation structure of tropical cyclones, their diurnal cycle, the evolution of the upper-level warm core and associated intensity changes, and the occurrence of intense precipitation cores or convective bursts. Moreover, the satellite data can be assimilated into mesoscale and global numerical prediction models.

A Category 5 equivalent, extremely severe cyclonic storm named Mocha had hit the Myanmar-Bangladesh coasts on 14 May 2023. It had winds of the order of 270 km/hr. Figure 8.1.2.1 is a Copernicus Sentinel-3 image of cyclone Mocha taken on 13 May 2023. Figure 8.1.2.2 is its TROPICS image of the same date showing clearly the rain bands that are already inside Myanmar while the cyclone eye is still over the Bay of Bengal.

Status updates on the TROPICS mission are available at ttps://science.nasa.gov/mission/tropics/

Figure 8.1.2.1 Copernicus Sentinel-3 image of cyclone Mocha over the Bay of Bengal on 13 May 2023.

(Source: https://www.esa.int/ESA_Multimedia/Images)

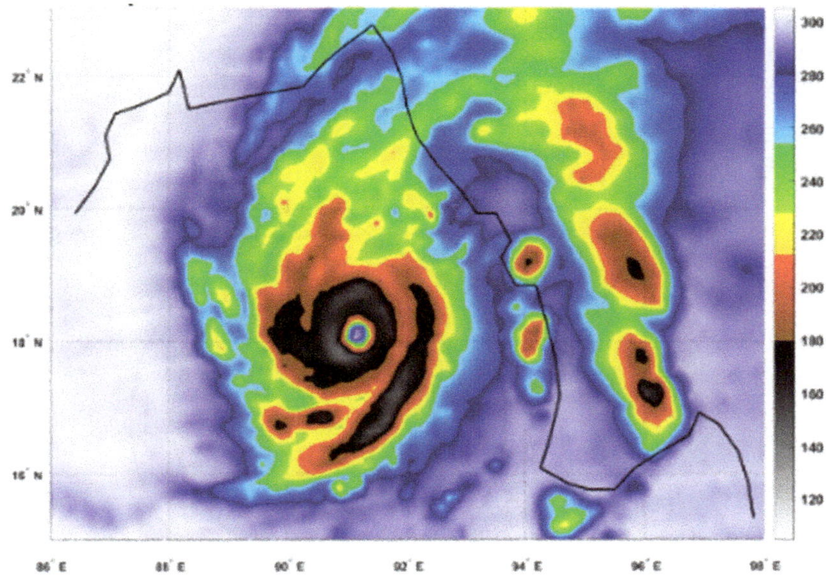

Figure 8.1.2.2 TROPICS 204.8 GHz image of cyclone Mocha over the Bay of Bengal on 13 May 2023.

(Source: https://cimss.ssec.wisc.edu)

8.1.3 GPS Radio Occultation Soundings

The Global Positioning System (GPS) is a satellite-based navigation system that provides location and time information anywhere on earth, as long as there is an unobstructed line of sight to four or more GPS satellites. Operated by the U.S., GPS consists of a constellation of 32 small satellites all moving in a circular orbit at an altitude of 20,200 km radius with a 55° inclination and a period of revolution of 12 hr around the earth. However, while their orbital characteristics are identical, the satellites move in six different orbital planes. Each GPS satellite broadcasts signals in two L-band frequencies called L1 and L2, at 1.57542 GHz or 19 cm wavelength, and 1.22760 GHz or 24.4 cm wavelength, respectively.

An application of great value is in an area that the original GPS concept was never intended to serve. This has come to be known as the GPS Radio Occultation (RO) technique which is now being used to retrieve temperature and humidity profiles in the atmosphere (Anthes et al 2008). The GPS RO technique is based on the principle that radio waves while passing through the atmosphere undergo refraction, the nature of which is in turn determined by the gradients of temperature, water vapour and electron density. Refraction effects are most pronounced when the radio wave traverses a long atmospheric limb path, and measurements for a series of such paths at different tangent heights contain information on the vertical profile of refractivity. If the refraction effects can be measured, then the atmospheric temperature and humidity elements that produced them can be retrieved.

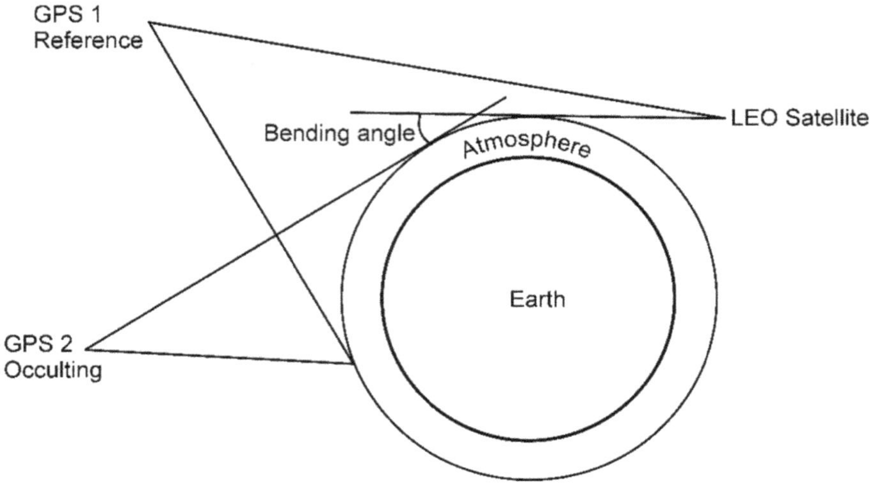

Figure 8.1.3.1 Schematic of the GPS radio occultation technique.

In practice, the GPS radio occultation technique involves precise L-band dual-frequency phase measurements by a GPS receiver placed on a low earth orbiting (LEO) satellite which tracks a setting or rising GPS satellite on the horizon (Figure 8.1.3.1). The GPS receiver has to be suitably equipped to detect amplitude, pseudo-range and phase measurements for each of these signals. The LEO satellite uses the GPS signals both to determine its own precise location and to make signal occultation measurements. As the position and velocity of both the LEO and GPS satellites are known, the small increase in the phase path due to the atmospheric refraction during the occultation event can be derived. This information is then converted to the atmospheric bending angles. Assuming a spherically symmetric atmosphere, vertical profiles of refractive index can be determined and converted to lower tropospheric soundings of atmospheric parameters such as pressure, temperature and indirectly water vapour.

The main advantages of the GPS soundings are global coverage, high vertical resolution and all-weather capability combined with high accuracy which make them ideally suited for various applications in atmospheric and ionospheric research, weather forecasting and climate studies. The practical feasibility of the GPS sounding technique was proven with an experimental satellite named CHAMP (CHAllenging Minisatellite Payload) launched in 2000 (Wickert et al 2001), and again with COSMIC (Taiwan-U.S.

Constellation Observing System for Meteorology, Ionosphere, and Climate) launched in 2006. COSMIC-2, a follow-up mission launched in 2019 with 6 satellites is currently operational, providing valuable radio occultation soundings for weather forecasting and climate monitoring.

GPS is part of a larger system called GNSS (Global Navigation Satellite Systems), which includes Russia's GLONASS, Europe's Galileo, and China's BeiDou. All of them are being used for RO sounding purposes. COSMIC yields about 5000 soundings daily. Besides that, private enterprises such as Spire and GeoOptics which operate large constellations of small satellites, are also involved, altogether producing an estimated total of 20,000 RO soundings a day.

8.1.4 Wildfires

Wildfires, particularly forest fires, are largely a natural phenomenon. In India, forest fires occur typically in the hot summer months of March to May and again in the post-monsoon dry months of November and December, posing a threat to the environment. However, in recent times, hotter and drier conditions caused by global warming are resulting in an increase in wildfire activity in many parts of the world. More organic dry matter is becoming available for burning, and higher temperatures help to start and spread the fires. As against this, forest fires can themselves contribute to global warming, as they release large amounts of stored carbon dioxide and other greenhouse gases into the atmosphere.

Not all wildfires are natural and many are anthropogenic, like post-harvest stubble burning in Punjab and Haryana in north India which has drawn increasing attention due to its adverse impact on regional air quality. As wildfires emit light and heat and are widespread, remote sensing of wildfires has traditionally been carried out using VIS and IR radiation data received from high resolution instruments like MODIS on NASA's Terra and Aqua satellites and VIIRS on NOAA satellites (Giglio et al 2016). Such data can be used to detect flames and smoke, to map the area on fire, and to follow its spreading making it possible to take disaster relief measures.

In a recent study (Singh et al 2025) data from the Indian satellite INSAT-3DS has been used for fire detection and monitoring. They have developed an algorithm that uses the MW and TIR bands of the INSAT-3DS imager, which offers observations at a 4 km spatial resolution every 30 minutes, as against the twice daily scans of orbiting satellites. It employs a contextual thresholding approach enhanced by optimized diurnal thresholds, a glint angle rejection scheme, and adaptive background pixel selection. The algorithm's efficacy is demonstrated through its successful detection of late-evening agricultural fires in Punjab and Haryana, as well as its ability to capture known fire incidents.

8.1.5 FireSat

The FireSat satellite network is a novel proposition of the Earth Fire Alliance (EFA) which was founded in 2024. EFA is a global, community-led, non-profit coalition that aims to deliver near real-time information about fires, particularly wildfires, anywhere on earth. It has many partners like Google Research, the Moore Foundation, and various wildfire authorities. FireSat is designed to enhance early wildfire detection and tracking, ultimately helping to mitigate the devastating impacts of wildfires.

Some information about the FireSat mission is posted on the EFA web page at https://www.earthfirealliance.org/, but full details are not yet available However, it is understood to be a low earth orbit satellite carrying a 6-channel multispectral radiometer which is capable of simultaneous imaging across VIS, NIR, SWIR, MWIR and TIR bands. It can detect fires over areas of the size 5 x 5 m, significantly smaller than what existing systems can detect. The observation swath width is 1,500 km. This multichannel data will be processed to derive heat signatures of wildfire development and progression.

A U.S.-based aerospace startup company Muon Space launched EFA's first FireSat satellite on SpaceX's Transporter-13 mission in March 2025, and it shared the first images taken by the spacecraft in July 2025. Eventually, FireSat will be a constellation of as many as 50 small satellites that would provide global information on wildfires every 20 min. AI technology will be used to analyze the satellite images, compare them with historical data and local conditions, and identify active fires, while also reducing false positives. It will trigger alerts for first responders and other stakeholders. More information is available at the Google Research web page https://sites.research.google/gr/wildfires/firesat/.

8.2 Lightning

Lightning is a phenomenon that is far deadlier than other hazards associated with a thunderstorm like hail, squalls and heavy rain. Hundreds of people lose their life every year across India because of lightning strikes particularly in the pre-monsoon summer months. Meteorological satellites are able to monitor thunderstorms but not lightning per se.

By definition, lightning is a high-energy luminous electrical discharge from a cumulonimbus (Cb) cloud to the ground which is accompanied by thunder. Electrical discharges can also occur in the atmosphere in two other ways. Intra-cloud lightning occurs within a Cb cloud and inter-cloud lightning is between two different Cb clouds.

The top of the Cb cloud develops a positive electric charge while its base has a negative charge. As air is a poor conductor of electricity, positive charge accumulates on all objects below the Cb cloud, and tends to concentrate on the highest object in the area. When the potential difference between this positive charge and the negative charge on the cloud becomes large enough to overcome the electrical resistance of the air, lightning occurs. The heavy electrical discharge of 100 million volts leaves the air heated to about 30,000 °C in a few milliseconds. This generates shock waves whose propagation is heard as thunder.

Since lightning can occur only when there is a thundercloud, data on lightning occurrence can serve as proxy for convective precipitation. Lightning also plays a role in atmospheric chemistry as it produces NOx in and around Cb clouds. This is carried to the upper troposphere by the convective updrafts and leads to an increase in the ozone concentrations at those levels.

The TRMM satellite launched in 1987, carried an instrument called the Lightning Imaging Sensor (LIS). This was used to detect lightning in all its three forms, cloud-to-cloud, intra-cloud, and cloud-to-ground, and it was possible to map out their frequency and distribution over the tropical regions. LIS consisted of a staring imager optimized to locate and detect lightning with storm-scale resolution (4 - 7 km) over a large region (600 × 600 km) of the earth's surface. In spite of the fact that the TRMM satellite moved at a speed of 7 km/sec, LIS had an observation time of 90 seconds over any place, which was long enough to estimate the flashing rate of most storms.

An instrument called the NASA Optical Transient Detector (OTD) was launched in April 1995 aboard the MicroLab-1 satellite into a 70° inclination orbit at an altitude of 740 km. At any given instant it viewed a 1300 × 1300 km region of the earth with a spatial resolution of 10 km.

Another Lightning Imaging Sensor (LIS) was flown aboard the International Space Station (ISS) to detect lightning activity across the globe. Its expected life was just two years but it worked for almost seven years up to November 2023. The ISS orbit enabled observations to be made over the mid-latitudes from 55°N to 55°S. Its detection efficiency was about 60 percent.

Because of the small dwell time of orbiting satellites over a given place and the long interval between their successive passes, a Geostationary Lightning Mapper (GLM) is being flown on the current GOES-R series of satellites. This is a single-channel, NIR (777 nm) staring CCD imager with a 10 km spatial resolution and a 20 sec temporal resolution. The spatial resolution is 8 km at nadir and 14 km at the edge. GLM can detect both daytime and nighttime lightning events making it very useful in tornado and other severe storm warnings.

Other geostationary satellites like Meteosat Third Generation (MTG) and Fengyun-4A also have a Lightning Imager (LI). INSAT-3D does not have a lightning sensor but efforts have been made to use its products like OLR, temperature and winds to identify the precursory signatures of the lightning occurrences with sufficient lead time (Goenka et al 2024, Rakshit et al 2024). INSAT-3D cloud top brightness temperatures are also being coupled with concurrent data from the Indian Lightning Location Network (ILLN), operated by the Indian Institute of Tropical Meteorology (IITM). A description of the methodology has been given by Mohapatra et al (2021) and Biswasharma et al (2025).

8.3 Planktons, Aerosols and Cloud Properties

The colour of the ocean is determined by the interaction of sunlight with substances or particles present in it such as chlorophyll which is a green pigment found in most phytoplankton species. Ocean colour is therefore a visual indicator of its phytoplankton content. Satellite observations of ocean colour have been extremely helpful for assessing the global distribution and abundance of phytoplankton, tiny plants and algae that sustain the marine food ecosystem. The first ocean colour satellite, Coastal Zone Colour Scanner (CZCS) was launched by the U.S. back in 1978. Many other ocean colour observing satellites have been launched thereafter like India's Oceansat satellites. These have carried radiometers sensitive to several discrete wavelength bands.

8.3.1 PACE

PACE is the latest satellite in this genre and by far the most advanced. PACE is an acronym for NASA's Plankton, Aerosol, Cloud, and ocean Ecosystem mission and it was launched in February 2024 (Werdell et al 2024). It is expected to achieve two science goals that are much broader than those of its predecessors:

(a) to extend key systematic ocean colour, aerosol, and cloud data records for earth system and climate studies, and

(b) to address new and emerging science questions using its advanced instruments, surpassing the capabilities of previous and current missions.

PACE will significantly improve the ability to understand the earth's changing marine ecosystems, manage natural resources such as fisheries, and identify harmful algal blooms. However, PACE is not limited to ocean colour monitoring. It is also simultaneously observing aerosols and clouds.

Aerosols are tiny airborne particles like dust, pollen, smoke or haze that absorb and scatter the incoming radiation. They directly alter the amount of solar energy reaching the earth and also indirectly through their interaction with clouds. The nature and influence of aerosols have not yet been fully understood and they are regarded as the largest source of uncertainty in quantifying the radiative forcing of climate change. PACE is measuring the size of cloud drops and ice crystals, and thereby improving the ability to differentiate between ice and water clouds.

The ocean and atmosphere are directly connected and keep exchanging energy, water, nutrients, gases, aerosols, and pollutants. Aerosols, clouds and phytoplankton can also affect one another. Certain aerosols may contribute to plankton blooms, while some plankton produce substances that act as seeds for cloud formation. PACE is designed to look at the ocean and the atmosphere and their interactions, all together.

The primary sensor of PACE is the Ocean Color Instrument (OCI). It is a hyperspectral radiometer that views a very broad wavelength range from the UV (350-400 nm), across VIS (400-700 nm), NIR (700-890 nm) bands to the SWIR (2250 nm), all at a nominal spatial resolution of 1 km. OCI measures the 345-890 nm region continuously with a resolution of 5 nm, while the SWIR wavelengths are measured in seven discrete bands (Remer et al 2019).

PACE also has two polarimetric radiometers to measure the polarization states of the UV to SWIR radiation at various angles. The Hyper-Angular Rainbow Polarimeter-2 (HARP2) provides atmospheric correction over the footprint of the OCI. HARP2 is designed to collect data over a wide field of view in the cross-track direction allowing for global coverage in about two days, as well as an even wider field of view in the along-track direction providing measurements over a wide range of scattering angles (Martins et al 2024).

The other instrument on PACE is SPEXone, which is a multi-angle polarimeter. It measures the intensity, degree of linear polarization and the angle of linear polarization of sunlight reflected back from the atmosphere, land surface and ocean. SPEXone has a compact three-mirror segmented telescope assembly to gather light from various angles and direct it towards a common entrance slit of the spectrometer (Fu et al 2025).

8.3.2 EarthCARE

The EarthCARE (Cloud, Aerosol and Radiation Explorer) mission is a joint venture of ESA and JAXA. The EarthCARE satellite can be said to be a great leap forward towards integrated remote sensing of clouds, aerosols, precipitation, and radiation. It had a long course of design and development (Illingworth et al 2015) and was launched in May 2024. The objective of this mission is to advance our understanding of the role that clouds and aerosols play in reflecting incident solar radiation back into space and trapping infrared radiation emitted from the earth's surface.

While the NASA A-Train was an array of several satellites flying in a formation mode, each carrying a payload for a different application, EarthCARE places all their individual instruments in a single satellite. The EarthCARE orbit has an altitude as low as 393 km, an inclination of 97° and a repeat cycle 25 days. EarthCARE has four main payloads:

The Cloud Profiling Radar (CPR) is a high power 94 GHz radar with Doppler capability. It measures the vertical velocity of cloud particles and their microscopic and macroscopic properties. Its microwave pulses can penetrate deep into the lower cloud layers that are not visible to optical instruments.

The Atmospheric Lidar (ATLID) is a high spectral resolution atmospheric backscatter lidar instrument that detects cloud boundaries and profiles optically thin clouds and aerosols. ATLID makes measurements in a direction close to the nadir along the satellite track, with a vertical resolution of about 103 m from ground upwards to 20 km altitude.

The Multi Spectral Imager (MSI) is a 7-band, push-broom scanner that provides images at 500 m ground sampling distance over a 150 km wide swath, which is offset from nadir pointing in order to minimize sun glint.

The Broad Band Radiometer (BBR) is the fourth instrument. While CPR and ATLID are used to measure cloud and aerosol profiles, BBR measures the co-located reflected solar and emitted thermal radiation. From the unfiltered BBR radiance measurements, broad band fluxes at the top of the atmosphere are derived.

In the course of its mission, EarthCARE is expected to yield the following data:

(a) Vertical profiles of natural and anthropogenic aerosols on a global scale, their radiative properties and interaction with clouds
(b) Vertical distributions of atmospheric liquid water and ice on a global scale, their transport by clouds and their radiative impact
(c) Cloud distribution, cloud-precipitation interactions and the characteristics of vertical motions within clouds
(d) Profiles of atmospheric radiative heating and cooling through the combination of the retrieved aerosol and cloud properties

8.3.3 AOS

NASA's PACE satellite can be regarded as a forerunner to its highly ambitious and futuristic Atmospheric Observing System (AOS) mission in several important ways, particularly regarding its contributions to aerosol and cloud studies (Bemdt et al 2023). Four AOS satellites will soon be in space, working together to make measurements of aerosols, clouds, convection and precipitation. They will be using a variety of instruments, both active and passive, in all wavelengths from microwave to UV, providing a comprehensive view of how atmospheric systems work and interact with each other.

The AOS mission has these primary objectives:

(a) to investigate vertical air motion within clouds,

(b) to understand how ice and water move vertically within clouds and how these motions relate to aerosols,

(c) to understand how aerosols: influence cloud formation and precipitation patterns,

(d) to study the dynamics of convective storms and the properties of different types of clouds in low, high, convective, and mixed phases,

(e) to examine how clouds and aerosols interact to influence the energy and water cycles.

All this is expected to improve weather predictions, air quality forecasts and climate modelling.

AOS will have two major components named as AOS-Storm and AOS-Sky which are targeted to commence in July 2028 and December 2031, respectively. Each of them will have two satellites. The AOS-Storm satellite will be paired with the PMM (Precipitation Measuring Mission) satellite in a shared 55° inclined orbit. The AOS-Sky satellite will be paired with the HAWCsat (High altitude Aerosol Water vapor Clouds satellite) in a shared polar orbit.

AOS-Storm will have a Backscatter Lidar to measure attenuated aerosol and cloud backscatter at 1064 nm and 532 nm wavelengths. It will also have a multi-channel Microwave Radiometer to measure ice water path, precipitation and atmospheric humidity. The PMM satellite will have a Ku-band (13.6 GHz) Doppler Radar with a 255 km wide swath to measure precipitation and the associated liquid or ice particles. It will have a similar Microwave Radiometer.

AOS-Sky will have a Single-frequency Doppler Radar to measure cloud properties and light precipitation, as well as the associated liquid or ice particle motions along the radar line of sight. This too will have a Microwave Radiometer and a Backscatter Lidar of common design. In addition, it will have a Polarimeter and a Far Infrared (FIR) Imaging Radiometer to measure ice particle size, cloud top height, cloud longwave emission and water vapor profiles. The HAWCsat satellite will carry an Aerosol Limb Imager and a Water Vapor Limb Imager to measure scattered sunlight from the edge of the atmosphere to retrieve the concentration of water vapour at various levels of the stratosphere and upper troposphere. Status updates on the AOS mission would be available at https://aos.gsfc.nasa.gov.

8.4 Atmosphere, Ocean and Land Parameters

In 1998, member countries of the European Union had expressed the need for having their own earth observation system, which they initially called GMES (Global Monitoring for Environment and Security). Many years later, this programme became a reality and was renamed as Copernicus, after the 16th century astronomer and mathematician who proved that the earth was not the centre of the universe but that it revolved around the sun. With the launch of its first satellite, Sentinel-1A, in 2014, the Copernicus programme became operational and is being implemented by the European Space Agency (ESA) and Eumetsat (Peuch et al 2022).

All subsequent Copernicus satellites have been named as Sentinel, even though they differ widely in design and functionality. Some of the Sentinel satellites, like Sentinel-3A and -3B, are identical twins or pairs revolving in a common orbit but spaced 180° apart so as to double their revisit frequency. Some Sentinels are not independent satellites but only instruments that are mounted on other operational polar orbiting or geostationary satellites. For example, Sentinel-4 is a spectrometer carried by the MTG-S geostationary satellite, while Sentinel-5A would be an instrument aboard an MetOp orbiting satellite.

8.4.1 Copernicus and Sentinels

Copernicus is a multi-dimensional program serving the interests of several application areas like land, ocean, atmosphere, climate change, emergency response, and security. Sentinel-1B has completed its mission, 1A, 1C, 2A, 2C, 3A, 3B, 4A, 5P and 6 are operational, while 2D, 3C, 3D, 4B, 5A and 6B are in the pipeline. Status updates are available on the ESA Copernicus web page at https://www.esa.int/.

Sentinel-1A was launched in April 2014 in a sun-synchronous, 98° inclination, 693 km altitude orbit with a 99 min orbital period and 12 days repeat cycle. It had a 5.4 GHz C-band Synthetic Aperture Radar (C-SAR) for all-weather, day and night monitoring of land, ocean and ice. Applications enabled by C-SAR were land deformation and subsidence mapping, sea ice and glacier monitoring, oil spill detection, maritime surveillance and emergency response.

The Sentinel-2 series satellites were placed in a sun-synchronous orbit with 786 km altitude, 99° inclination, 101 min orbital period, and 5 days revisit time with two satellites phased 180° apart. They carry an optical instrument that samples 13 spectral bands, including: four bands at 10 m, six bands at 20 m, and three bands at 60 m spatial resolution. The orbital swath width is 290 km. The multispectral imaging is useful for monitoring of vegetation, soil, land use change, inland waterways and coastal areas.

The Sentinel-3 series was started in 2016. Sentinel-3A was placed in a sun-synchronous polar orbit at an altitude of 815 km and inclination of 99° with an orbital period of 101 minutes. It had a repeat cycle of 27 days. It had four instruments on board. The Ocean and Land Colour Instrument (OLCI) had 21 spectral bands with 300 m resolution for measuring ocean and land surface colour, optimized to reduce sun glint interference. OLCI also supported monitoring of the GHGs, ozone and aerosols. The Sea and Land Surface Temperature Radiometer (SLSTR) had 11 spectral bands with 500 m resolution for VIS/SWIR, and 1 km resolution for thermal IR for measuring sea and land surface temperatures with 0.3 °K accuracy, with dual-view scanning for atmospheric correction. It could also detect active fires using dedicated bands. The Synthetic Aperture Radar Altimeter (SRAL) was a dual-frequency Ku and C-band altimeter for measuring sea surface height, wave height, and topography at a resolution of 300 m along track. It was useful over oceans, sea ice, rivers, and lakes. A Microwave Radiometer (MWR) provided atmospheric corrections for SRAL and measured atmospheric water vapor content.

The Sentinel-6A satellite was launched in November 2020 in a sun-synchronous orbit at 1,336 km altitude with a 10-day revisit cycle and is currently operational. It has a radar altimeter that measures the sea surface height and it is carrying further the legacy of the Poseidon-4 radar altimeter. Sentinel-6 also has on board an Advanced Microwave Radiometer (AMR-C) for atmospheric water vapour correction and a GNSS-RO for retrievals of vertical temperature profiles in the troposphere and stratosphere. Its twin satellite, Sentinel-6B, is scheduled for launch in November 2025, ensuring continuity of high-precision ocean altimetry through at least 2030.

The Sentinel-5P precursor satellite was launched in 2017 in a sun-synchronous low earth orbit at 820 km altitude and 99° inclination with a repeat cycle of 29 days. It carries a Tropospheric Monitoring Instrument (TROPOMI), developed in the Netherlands. This is a multi-spectral imaging spectrometer that is sensitive to UV and VIS (270–500 nm), NIR (675–775 nm) and SWIR (2305–2385 nm) bands of the spectrum. It uses a push-broom configuration to scan a wide swath of 2,600 km of the earth's surface daily with a high spatial resolution of 3.5 ×7 km.

By measuring the radiation scattered by the earth's surface and the atmosphere, TROPOMI is able to detect the unique fingerprints of various trace gases in the atmosphere such as nitrogen dioxide, sulphur dioxide, ozone, carbon monoxide and methane, and also aerosols, for climate and air quality studies.

Sentinel-4A performs a similar function as TROPOMI but it is not an independent satellite like Sentinel-5P. Sentinel-4A is the name given to a UVN Spectrometer which is mounted on the Meteosat Third Generation Sounder

(MTG-S1) satellite which was launched in July 2025. It measures radiation in UV (305–400 nm), VIS (400–500 nm), and NIR (750–775 nm) bands. Like TROPOMI, this data is used to retrieve the concentrations of trace gases in the atmosphere. Since the host satellite is a geostationary satellite, the geographical coverage is limited, but the earth's atmosphere is scanned at hourly intervals with a spatial resolution of 8 × 8 km. An identical instrument is planned to be flown on MTG-S2 satellite in the year 2030 as Sentinel-4B.

8.5 Ecosystems, Earth's Crust and Cryosphere

A conventional ground-based imaging radar emits an electromagnetic signal towards an object and receives back the reflected signal or echo. Measurements of the strength of the echo and the time taken by it to return help to interpret the nature of the object and determine its position. For the radar beam to be sharply focused, the radar antenna has to be large. On the other hand, if a radar is required to be placed on a satellite to observe the earth's surface, the antenna has to be necessarily small. Hence the signal beam gets thinly spread out as it reaches the earth's surface and the backscatter is weak.

The SAR is a technique that overcomes this limitation and produces high-resolution images of the earth's surface. It takes advantage of the situation that the radar on board a satellite is not stationary but in motion. As it moves, it collects multiple echoes from the same target thus synthesizing a huge antenna that could virtually be a thousand times larger than the real one. Images of the earth's surface with a resolution as miniscule as 1 cm can thus be produced.

8.5.1 NISAR

The NASA ISRO Synthetic Aperture Radar (NISAR), as obvious from its name, is a joint NASA/ISRO mission that aims at global high resolution measurements of land surface changes using advanced radar imaging. While exploratory work had begun way back in 2007, the NISAR concept was formalized in 2014 and underwent a long process of development (Rosen et al 2024). NISAR was successfully launched on 30 July 2025 aboard an ISRO Geosynchronous Launch Vehicle Mark II rocket from the Satish Dhawan Space Centre in Sriharikota, India. This Indo-U.S. collaborative project is said to have had a total outlay close to 1.5 billion U.S. dollars, and NISAR is ranked supposedly as the most expensive earth observation satellite ever built.

NISAR is unique in the sense that it carries a payload of not just one synthetic aperture radar (SAR) but two (Figure 8.5.1). The NASA radar operates in the L-band (24 cm wavelength) and the ISRO radar operates in the S-band (9.4 cm wavelength). NASA contributed to the communications and data systems. ISRO provided the spacecraft bus, the S-band radar, the launch vehicle and was responsible for the launch itself.

The three-axis stabilized satellite was launched into a sun-synchronous near polar orbit having 98.4° inclination and 748 km altitude with a swath of 240 km. The orbit is designed so as to complete the view the of the entire earth over a period of 12 days. The spacecraft is expected to have a life of 5 years.

The NISAR mission envisages the utilization of the data in three domains of global importance and another three of special interest to India (Rosen et al 2025).

1. *Ecosystems*: monitoring of agricultural crops and crop biomass; soil moisture; forest inventory, biomass change, and carbon sequestration; mangroves and wet lands; and tree-line ecotone in alpine vegetation
2. *Solid Earth*: risk assessment and management through monitoring co-seismic, post-seismic, and inter-seismic deformation; volcanic deformation; landslides; and aquifer systems for groundwater flow, storage, and management
3. *Cryosphere*: monitoring the dynamics of the polar ice sheet (ice shelf, land ice velocity, and ice discharge to the ocean); sea ice types, thickness, and motion for ship safety; and Himalayan snow and glacier dynamics for hydrological assessments
4. *Oceanography*: retrieval of ocean physical parameters such as surface wind speed, wave spectra, and coastal bathymetry for shipping, improvements in model simulations, cyclone monitoring, and storm surge
5. *Coastal processes*: monitoring coastal erosion processes, near-shore dynamics, and coastal subsidence; assessment of coastal vulnerability to sea level rise and shoreline change; and monitoring coastal critical habitats.
6. *Disaster response*: mapping and monitoring of floods, oil spills, and earthquake damage and monitoring the damage due to forest fires and impact of extreme weather events such as tropical cyclones

The land surface changes detected by NISAR over the course of multiple orbits, would be interpreted in a variety of practical application. It will be possible to identify landslide-prone areas, and seismically active zones. Measurements of ground deformation along faults before an earthquake occurs, from the earthquake itself, and in the time following, will help understand where and why earthquakes occur. NISAR will monitor volcanic eruptions, flow rates of glaciers and ice sheets and sea ice over Arctic and Antarctic. NISAR will provide maps of surface soil moisture globally every 6 to 12 days at the spatial scale of individual farm fields

NISAR will go through a post-launch commissioning phase of 90 days after which early data is expected to be released. NISAR data policy is to make the data freely available to the global community. Status updates will be available at https://science.nasa.gov/mission/nisar/.

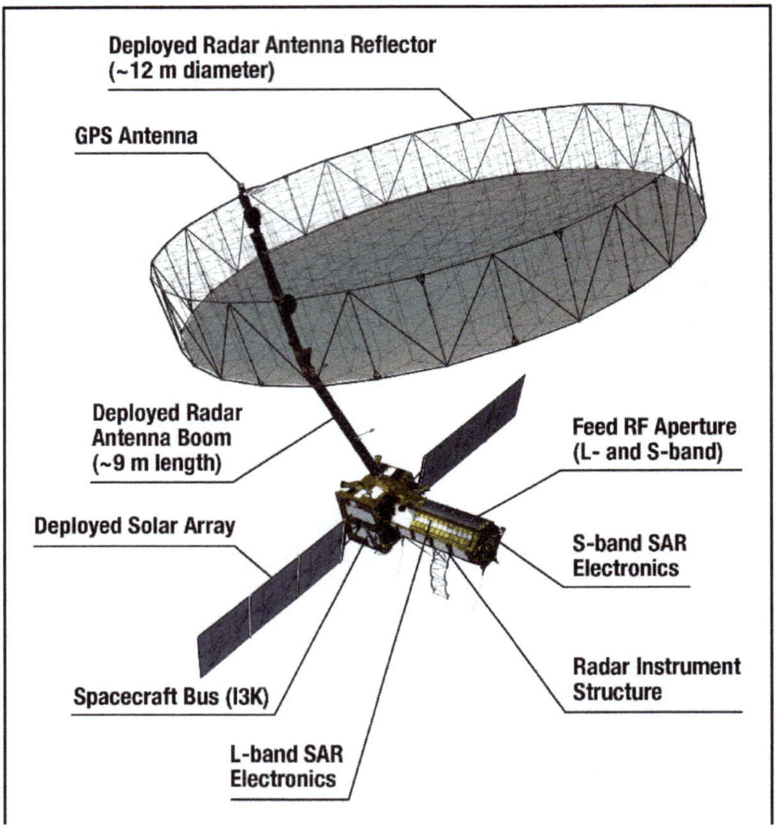

Figure 8.5.1.1 NISAR spacecraft schematic.
(Source: https://science.nasa.gov/mission/nisar/)

8.6 The Future Lies in AI

Besides all the trends and possibilities described above which are hopefully going to be realized, satellite meteorology is now poised at a new threshold, waiting expectantly for the advent of Artificial Intelligence and Machine Learning. AI will enable satellites to process data on board the satellite itself without losing precious time in transmitting it to the ground. We will see AI transforming satellites from observing platforms into intelligent systems. This would be critical for real time forecasting of severe storms, lightning strikes, wildfires and flash floods.

Already, claims are being made of AI-based weather forecasts beating NWP forecasts in speed and accuracy (Soliman 2024). Conventional forecasts, including those from model ensembles, are based on mathematical models that simulate the laws of physics and dynamics governing the earth's atmosphere and oceans. Powerful supercomputers are needed for solving these complex mathematical equations, and they have to be provided with satellite and other data as initial conditions. AI models in contrast, are data driven, and use historical as well as concurrent data. Private enterprises are engaged in launching large numbers of small satellites with radars and microwave sounders and with onboard processing to save time and effort. Cutting edge technology companies are reshaping weather forecasts, to make them hyper-local, and in truly real-time.

AI is certain to reshape satellite meteorology and offer deeper insights, enable faster decision making, and allow wider accessibility. This synergy is key to addressing significant issues related to climate change, disaster resilience, and sustainable development.

8.7 In Conclusion

The ultimate goal of the weather and climate community is to achieve a completely interactive simulation of the dynamical, radiative, and chemical processes taking place in the integrated atmosphere-ocean-land system. There is no doubt that satellite meteorology has already made a significant contribution towards achieving this goal, and will continue to do so in future.

"What is faith?" is a question posed in the Bible, and the answer that follows says, "Faith gives substance to our hopes and makes us certain of realities we do not see" (NEB 1961). Except for the changing colour of the sky, and the beauty and fury of clouds and rain, there is little else in the atmosphere that we can actually see. It is satellite meteorology that, in the manner of faith, makes us certain of the realities of the atmosphere that we are not able to see with our own vision.

8.8 References

Anthes R. A. et al, 2008, "The COSMIC/FORMOSAT-3 mission: Early results", *Bull. Amer. Meteor. Soc.*, 89, 313-333.

Berndt E. B. et al, 2023, "From NASA's EOS to ESO: advancing applications of the future Atmosphere Observing (AOS) mission", *103rd American Meteorological Society (AMS) Annual Meeting.*

Biswasharma R. et al, 2025, "Assessment of the Indian Lightning Location Network (ILLN) using ground-based and satellite observations." *Atmospheric Res.*, 320, 108069.

Blackwell W. J. et al, 2018, “An overview of the TROPICS NASA earth venture mission”, *Quart. J. Royal Meteor. Soc.*, 144, 16-26.

Carreno-Luengo H. et al 2021, "The CYGNSS mission: on-going science team investigations", *Remote Sensing,* 13, 1814.

Fu G. et al, 2025, "Aerosol retrievals from SPEXone on the NASA PACE mission: first results and validation." *Geophys. Res. Letters 52.4: e2024GL113525.*

Giglio L, Schroeder W., Justice C. O., 2016, “The collection 6 MODIS active fire detection algorithm and fire products”, *Remote Sensing Environ.*, 178, 31–41.

Goenka R., Taori A., Rao G. S. and Chauhan P., 2025, “Leveraging INSAT-3D Indian geostationary satellite for advanced lightning detection and analysis”, *Geophys. Res. Letters*, 52, e2024GL112764.

Illingworth A. J. et al, 2015, “The EarthCARE satellite: the next step forward in global measurements of clouds, aerosols, precipitation and radiation”, *Bull. Amer. Meteor. Soc.,* 96, 1311-1332.

Martins J. V. et al, 2024, "First results and on-orbit performance of the Hyper-Angular Rainbow Polarimeter (HARP2) on the PACE satellite", *Sensors, Systems, and Next-Generation Satellites XXVIII. Vol. 13192. SPIE.*

Mohapatra M. et al, 2021, “INSAT-3DR rapid scan operations for weather monitoring over India”, *Current Sci.*, 120, 1026-1034.

Peral E. et al, 2015, “Raincube: A proposed constellation of precipitation profiling radars in CubeSat”, *Geosci. Remote Sensing Symp. (IGARSS), 2015 IEEE International*, 1261-1264.

Peuch V. H. et al, 2022, “The Copernicus atmosphere monitoring service: from research to operations”, *Bulletin American Meteor. Soc.*, 103, 2650-2668.

Rakshit G., Mitra A. K. and Krishnan K. S., 2024, “Multitechnique investigation of the prevailing atmospheric parameters for the Odisha lightning catastrophe on 2 September 2023: a case study”, *J. Earth System Sci.*, 134, 2.

Remer L. A. et al, 2019, "Retrieving aerosol characteristics from the PACE mission, Part 1: Ocean Color Instrument." *Frontiers in Earth Sci.,* 7, 152.

Rosen P. A. et al, 2024, "The NASA-ISRO SAR mission leading up to launch", *IEEE India Geoscience and Remote Sensing Symposium (Ingres)..*

Rosen P. A. et al, 2025, "The NASA-ISRO SAR mission: a summary" *IEEE Geoscience and Remote Sensing Magazine*, doi 10.1109/MGRS.2025.3578258.

Singh N. et al, 2025, “Active fire detection using Indian geostationary satellite INSAT-3DS”, *Remote Sensing. Earth System Sci.,* doi.org/10.1007/s41976-025-00233-4

Soliman A., 2024, "DeepMind AI weather forecaster beats world-class system", *Nature*, 636, 282-283.

Werdell P. J. et al, 2024, "Life after launch: a snapshot of the first six months of NASA’s Plankton, Aerosol, Cloud, ocean Ecosystem (PACE) mission", *Sensors, Systems, and Next-Generation Satellites XXVIII,* Vol. 13192. SPIE.

Wickert J. et al, 2001, “Atmospheric sounding by GPS radio occultation: first results from CHAMP”, *Geophys. Res. Letters*, 28, 3263.

Subject Index

R

S

W

X

www.ingramcontent.com/pod-product-compliance
Ingram Content Group UK Ltd.
Pitfield, Milton Keynes, MK11 3LW, UK
UKHW021350300726
14134UKWH00026B/470